AF386563

INFORMATION SHARING FRAMEWORK
FOR REGULATORY REVIEWS
OF ADVANCED REACTORS

The following States are Members of the International Atomic Energy Agency:

AFGHANISTAN	GEORGIA	PAKISTAN
ALBANIA	GERMANY	PALAU
ALGERIA	GHANA	PANAMA
ANGOLA	GREECE	PAPUA NEW GUINEA
ANTIGUA AND BARBUDA	GRENADA	PARAGUAY
ARGENTINA	GUATEMALA	PERU
ARMENIA	GUINEA	PHILIPPINES
AUSTRALIA	GUYANA	POLAND
AUSTRIA	HAITI	PORTUGAL
AZERBAIJAN	HOLY SEE	QATAR
BAHAMAS, THE	HONDURAS	REPUBLIC OF MOLDOVA
BAHRAIN	HUNGARY	ROMANIA
BANGLADESH	ICELAND	RUSSIAN FEDERATION
BARBADOS	INDIA	RWANDA
BELARUS	INDONESIA	SAINT KITTS AND NEVIS
BELGIUM	IRAN, ISLAMIC REPUBLIC OF	SAINT LUCIA
BELIZE	IRAQ	SAINT VINCENT AND
BENIN	IRELAND	THE GRENADINES
BOLIVIA, PLURINATIONAL	ISRAEL	SAMOA
STATE OF	ITALY	SAN MARINO
BOSNIA AND HERZEGOVINA	JAMAICA	SAUDI ARABIA
BOTSWANA	JAPAN	SENEGAL
BRAZIL	JORDAN	SERBIA
BRUNEI DARUSSALAM	KAZAKHSTAN	SEYCHELLES
BULGARIA	KENYA	SIERRA LEONE
BURKINA FASO	KOREA, REPUBLIC OF	SINGAPORE
BURUNDI	KUWAIT	SLOVAKIA
CABO VERDE	KYRGYZSTAN	SLOVENIA
CAMBODIA	LAO PEOPLE'S DEMOCRATIC	SOMALIA
CAMEROON	REPUBLIC	SOUTH AFRICA
CANADA	LATVIA	SPAIN
CENTRAL AFRICAN	LEBANON	SRI LANKA
REPUBLIC	LESOTHO	SUDAN
CHAD	LIBERIA	SWEDEN
CHILE	LIBYA	SWITZERLAND
CHINA	LIECHTENSTEIN	SYRIAN ARAB REPUBLIC
COLOMBIA	LITHUANIA	TAJIKISTAN
COMOROS	LUXEMBOURG	THAILAND
CONGO	MADAGASCAR	TOGO
COOK ISLANDS	MALAWI	TONGA
COSTA RICA	MALAYSIA	TRINIDAD AND TOBAGO
CÔTE D'IVOIRE	MALI	TUNISIA
CROATIA	MALTA	TÜRKİYE
CUBA	MARSHALL ISLANDS	TURKMENISTAN
CYPRUS	MAURITANIA	UGANDA
CZECH REPUBLIC	MAURITIUS	UKRAINE
DEMOCRATIC REPUBLIC	MEXICO	UNITED ARAB EMIRATES
OF THE CONGO	MONACO	UNITED KINGDOM OF
DENMARK	MONGOLIA	GREAT BRITAIN AND
DJIBOUTI	MONTENEGRO	NORTHERN IRELAND
DOMINICA	MOROCCO	UNITED REPUBLIC OF TANZANIA
DOMINICAN REPUBLIC	MOZAMBIQUE	UNITED STATES OF AMERICA
ECUADOR	MYANMAR	URUGUAY
EGYPT	NAMIBIA	UZBEKISTAN
EL SALVADOR	NEPAL	VANUATU
ERITREA	NETHERLANDS,	VENEZUELA, BOLIVARIAN
ESTONIA	KINGDOM OF THE	REPUBLIC OF
ESWATINI	NEW ZEALAND	VIET NAM
ETHIOPIA	NICARAGUA	YEMEN
FIJI	NIGER	ZAMBIA
FINLAND	NIGERIA	ZIMBABWE
FRANCE	NORTH MACEDONIA	
GABON	NORWAY	
GAMBIA, THE	OMAN	

The Agency's Statute was approved on 23 October 1956 by the Conference on the Statute of the IAEA held at United Nations Headquarters, New York; it entered into force on 29 July 1957. The Headquarters of the Agency are situated in Vienna. Its principal objective is "to accelerate and enlarge the contribution of atomic energy to peace, health and prosperity throughout the world".

IAEA-TECDOC-2105

INFORMATION SHARING FRAMEWORK FOR REGULATORY REVIEWS OF ADVANCED REACTORS

THE IAEA'S NUCLEAR HARMONIZATION AND STANDARDIZATION INITIATIVE

INTERNATIONAL ATOMIC ENERGY AGENCY

VIENNA, 2025

COPYRIGHT NOTICE

All IAEA scientific and technical publications are protected by the terms of the Universal Copyright Convention as adopted in 1952 (Geneva) and as revised in 1971 (Paris). The copyright has since been extended by the World Intellectual Property Organization (Geneva) to include electronic and virtual intellectual property. Permission may be required to use whole or parts of texts contained in IAEA publications in printed or electronic form. Please see www.iaea.org/publications/rights-and-permissions for more details. Enquiries may be addressed to:

Publishing Section
International Atomic Energy Agency
Vienna International Centre
PO Box 100
1400 Vienna, Austria
tel.: +43 1 2600 22529 or 22530
email: sales.publications@iaea.org
www.iaea.org/publications

For further information on this publication, please contact:

Regulatory Activities Section
International Atomic Energy Agency
Vienna International Centre
PO Box 100
1400 Vienna, Austria
Email: Official.Mail@iaea.org

© IAEA, 2025
Printed by the IAEA in Austria
December 2025
https://doi.org/10.61092/iaea.xiip-aot6

IAEA Library Cataloguing in Publication Data

Names: International Atomic Energy Agency.
Title: Information sharing framework for regulatory reviews of advanced reactors / International Atomic Energy Agency.
Description: Vienna : International Atomic Energy Agency, 2025. | Series: IAEA-TECDOC ISSN 1011-4289 ; no. 2105 | Includes bibliographical references.
Identifiers: IAEAL 25-01799 | ISBN 978-92-0-124025-5 (paperback : alk. paper) | ISBN 978-92-0-124125-2 (pdf)
Subjects: LCSH: Nuclear reactors — Information resources. | Nuclear reactors — Information technology. | Nuclear reactors — Information services. | Information networks..

FOREWORD

In recent years, there has been growing interest in the global deployment of standardized advanced nuclear reactors, including small modular reactors. This trend has been accompanied by an increase in regulatory reviews and has placed greater demands on regulatory resources. At the same time, the current differences in regulatory and industrial approaches among countries have made the standardization of reactor designs across national borders challenging.

To address these challenges, the IAEA launched the Nuclear Harmonization and Standardization Initiative (NHSI) in 2022 to support the effective global deployment of safe and secure advanced nuclear reactors. The initiative is structured in two interfacing tracks: one for technology holders and operators (the Industry Track) and one for regulators (the Regulatory Track).

The NHSI Industry Track aims to develop tools and industrial approaches for the effective large scale deployment of advanced reactors, with particular emphasis on small modular reactors. In parallel, the NHSI Regulatory Track aims to develop a global framework for the regulatory review of advanced reactors, also with particular attention to small modular reactors. The framework is intended to outline common regulatory requirements and establish a shared understanding of how to meet them; to enhance national reviews by enabling regulatory bodies to take maximum advantage of international efforts and the work of other regulatory bodies; and to enable the sharing of regulatory resources and the implementation of joint reviews, without introducing additional regulatory steps or increasing the duration of national licensing processes.

To develop this global framework, a clear, staged approach was envisaged for the NHSI Regulatory Track, with three distinct phases of work. The first phase, completed in 2024, focused on the development of processes and tools to promote cooperation in regulatory reviews and increase alignment in review outcomes. It is envisaged that the second phase will focus on implementing the processes and tools developed during the first phase, as well as on gathering feedback to improve cooperation processes and to map the regulatory differences among Member States. The final phase is planned to focus on assembling the elements necessary to establish the global framework for regulatory reviews based on the feedback collected, in addition to building on the identified regulatory requirement commonalities and launching targeted efforts to address the differences.

During the first phase, the members of the NHSI Regulatory Track, including regulatory bodies and industry representatives, collaborated through three dedicated working groups tasked with developing processes to enhance regulatory cooperation. Working Group 1 developed a framework to enable information sharing among regulatory bodies in order to facilitate cooperation in reviews of advanced reactors. Working Group 2 developed a multinational pre-licensing joint regulatory review process, in which a team of regulatory bodies conducts a design review against common requirements and reaches a joint decision. Working Group 3, led by the Small Modular Reactor Regulators' Forum, focused on identifying practical approaches for leveraging existing regulatory reviews and international collaboration in the regulatory review process.

This publication presents the framework developed by Working Group 1 to support information sharing among regulatory bodies and facilitate cooperation in reviews of advanced reactors. It outlines the information needed for a typical regulatory review, identifies potential barriers to cross-jurisdictional information exchange and proposes practical solutions to overcome these challenges. The publication also highlights examples of existing multinational collaborations that offer valuable insights and can inform effective approaches to regulatory cooperation.

This publication is primarily intended for regulatory bodies and technical support organizations, but it is also relevant to industry stakeholders, such as vendors that support licensees, because it provides insights into the information sharing process that can enhance communication and collaboration between regulatory bodies and the industry. By fostering a more transparent and consistent basis for the global review of advanced reactor designs, the framework supports the early identification and resolution of potential regulatory issues, thereby reducing risks and facilitating the approval process during subsequent national licensing stages.

The IAEA is grateful to all the experts who participated in the NHSI Regulatory Track Working Group 1 consultancy meetings organized for the preparation of this publication and/or contributed to its review. The IAEA officers responsible for this publication were P. Calle Vives and M. Ascic of the Division of Nuclear Installation Safety.

EDITORIAL NOTE

This publication has been prepared from the original material as submitted by the contributors and has not been edited by the editorial staff of the IAEA. The views expressed remain the responsibility of the contributors and do not necessarily represent the views of the IAEA or its Member States.

Guidance and recommendations provided here in relation to identified good practices represent expert opinion but are not made on the basis of a consensus of all Member States.

Neither the IAEA nor its Member States assume any responsibility for consequences which may arise from the use of this publication. This publication does not address questions of responsibility, legal or otherwise, for acts or omissions on the part of any person.

The use of particular designations of countries or territories does not imply any judgement by the publisher, the IAEA, as to the legal status of such countries or territories, of their authorities and institutions or of the delimitation of their boundaries.

The mention of names of specific companies or products (whether or not indicated as registered) does not imply any intention to infringe proprietary rights, nor should it be construed as an endorsement or recommendation on the part of the IAEA.

The authors are responsible for having obtained the necessary permission for the IAEA to reproduce, translate or use material from sources already protected by copyrights.

The IAEA has no responsibility for the persistence or accuracy of URLs for external or third party Internet web sites referred to in this publication and does not guarantee that any content on such web sites is, or will remain, accurate or appropriate.

CONTENTS

1. INTRODUCTION

Various advanced reactor designs and in particular small modular reactor (SMR) designs are being considered for deployment by countries around the world [1]. When deploying an SMR design to several States, it is advantageous if the design changes arising from differences among States' regulations are minimized. This might be facilitated by cooperation during regulatory reviews, including collaborative reviews [2] and joint reviews [3] in which regulatory bodies from different States work together, and by 'leveraging' (making best use of) reviews previously performed by other regulatory bodies [2]. Such cooperation between regulatory bodies has the potential to allow for easier international deployment of advanced reactors, both to countries with nuclear experience and to embarking countries, without compromising safety.

In accordance with IAEA Safety Standards Series No. SF-1, Fundamental Safety Principles [4], Principle 2, "The regulatory body must: …Have adequate legal authority, technical and managerial competence, and human and financial resources to fulfil its responsibilities." One of the responsibilities of the regulatory body in the framework of regulatory reviews is to assess the suitability of the supplied information for its intended purpose. Regulatory cooperation does not diminish the responsibility of the regulatory body to competently perform its duties.

Cooperation during regulatory reviews of advanced reactors can enhance national reviews, potentially reduce time and resources needed for the review, increase safety through a more thorough review and the sharing of good practices, and provide a flexible way for regulatory bodies to work together. Regulatory cooperation can also provide useful experience to both experienced regulatory bodies and regulatory bodies in embarking countries. Furthermore, regulatory cooperation can reduce the time and costs of licensing 'nth of a kind' reactor designs.

1.1. BACKGROUND

The Nuclear Harmonization and Standardization Initiative (NHSI) was established by the IAEA Director General in early 2022 in response to growing interest in advanced nuclear reactors. Under NHSI, governments, regulatory bodies, technical support organizations (TSOs), designers, technology owners, operating organizations and international organizations came together in a collaborative effort, consistent with their assigned roles and responsibilities, to harmonize and standardize regulatory and industrial approaches in support of the global deployment of safe and secure advanced nuclear reactors. NHSI consists of two tracks: an Industry Track and a Regulatory Track. This publication was developed under the Regulatory Track. It is applicable to any advanced reactor covered by the existing regulations of States, including SMRs.

The NHSI Regulatory Track supports the establishment of an international framework that will enable increased cooperation of regulatory bodies during advanced reactor reviews and during leveraging of regulatory reviews [2] and resources. This framework could be used as the basis for future regulatory harmonization in the licensing of new technologies. The key aspects of the NHSI Regulatory Track are:

- Minimizing repetition among regulatory reviews by different States;

- Minimizing the need for design changes arising from differences among regulations of States;

- Establishing a common basis for States' regulatory decisions while preserving States' sovereignty.

The approaches developed within this track are meant to enhance national reviews, enabling regulatory bodies to take maximum advantage of international work and efforts by other regulatory bodies. The implementation of these approaches is not expected to result in additional steps or increases in the duration of national licensing processes.

The approaches for regulatory cooperation developed within the NHSI Regulatory Track have focused on three types of cooperation:

- Collaborative review: an independent review against national requirements, discussing with other regulatory bodies but potentially reaching different decisions.

- Joint review: a team of regulatory bodies jointly review a design against common requirements and reaches a joint decisions.

- Leveraging of regulatory reviews: a review against national requirements with the use of other regulatory bodies' reviews.

The work of the NHSI Regulatory Track was divided into three topics, each of which were addressed by a different working group:

- Working Group 1 - Information sharing framework for regulatory reviews of advanced reactors (presented in this publication);

- Working Group 2 - Multilateral pre-licensing joint regulatory review (see Ref. [3]);

- Working Group 3 - Collaborative reviews and effective leveraging of regulatory reviews (see Ref. [2]).

1.2. OBJECTIVE

This publication aims to contribute to the establishment of a new international framework that will enable the sharing of information among regulatory bodies during advanced reactor pre-licensing and licensing reviews. Specifically, it describes the information that needs to be shared and potential impediments to this sharing. Moreover, the document presents possible solutions to overcome these impediments.

1.3. SCOPE

This publication covers the sharing of information that is associated with regulatory reviews during pre-licensing or licensing of advanced reactors, including SMRs. These reviews can be done by one or more regulatory bodies from a single State using the national framework or by several regulatory bodies from two or more States working together and using processes developed for the purpose by the States in question or that have been developed as part of NHSI. It is important to acknowledge that each State retains its national sovereignty over regulations, meaning that, while recognizing the benefits of the collaboration between regulatory bodies, the ultimate authority and implementation of regulatory functions and processes lie within the jurisdiction of each respective State. Therefore, not all aspects of collaboration between different regulatory bodies discussed in this publication may align perfectly with the specific regulatory framework of a particular State.

The principles outlined for information sharing in this publication apply to any type of reactor covered by the existing regulations of the States. This does not include the sharing of information on some aspects of the fuel design and the transportation of nuclear fuel that are associated with safeguards; where regulatory cooperation on the review of advanced reactors includes these aspects, additional steps (not described in this publication) will need to be devised and agreed for the sharing of information. The regulatory reviews discussed in this publication encompass all design and safety aspects that may be involved in pre-licensing or licensing processes.

In the context of international regulatory cooperation, potential conflicts of interest need to be carefully considered to ensure the integrity and objectivity of shared information and decision making processes. Conflicts of interest might arise when organizations or individuals involved in information sharing hold competing responsibilities or relationships that could influence regulatory outcomes. The issues related to conflict of interest are out of scope of this publication. Stakeholders in cooperative projects are encouraged to establish clear guidelines for identifying, managing and mitigating conflicts of interest within collaborative and joint review projects. By addressing these factors proactively, regulatory bodies can promote trust, transparency and impartiality in shared regulatory reviews.

1.4. STRUCTURE

This publication is organized into four sections and three annexes. Section 2 describes the nature of the information that needs to be shared among regulatory bodies when conducting joint reviews, collaborative reviews and when leveraging regulatory reviews. Section 3 then identifies potential impediments to such sharing of information. Section 4 consolidates and summarizes the key impediments identified in Section 3 and offers potential solutions to mitigate these challenges. Section 4 also describes some examples of collaborative reviews that have been successfully undertaken.

Annex I provides a proposed memorandum of cooperation (MoC) for consideration by States planning to share information and collaborate on regulatory reviews of advanced reactors. Annex II provides an overview of typical pre-licensing and licensing information needs to be considered by States. Annex III describes potential approaches for conducting a multinational review of design documents.

2. IDENTIFICATION OF INFORMATION NEEDS TO FACILITATE REGULATORY COOPERATION

This section identifies the nature of the information that needs to be shared for regulatory cooperation during the review of advanced reactors. Regulatory cooperation can take many forms, the specifics of which depend on the following characteristics, described in more detail in Annex III:

- Depending on the type of regulatory body, including internal or external TSOs participating in the review process;
- Depending on the number of participating States, regulatory bodies and TSOs (two or a larger group);
- Depending on participation of a body, network or organization as secretariat or for administration of review processes.

However, this publication report is concerned with the following specific forms of cooperation:

- **Collaborative reviews**: each participating regulatory body reviews all the information against its national requirements or against other agreed requirements, confers with other participants and reaches its own decisions. Further information about these reviews can be found in Ref. [2].

- **Joint reviews**: a team identified by participating regulatory bodies jointly reviews all the information against agreed requirements and comes to a joint decision. NHSI Working Group 2 has developed a process to perform multinational pre-licensing joint regulatory reviews. Further information about this process can be found in Ref. [3].

- **Leveraging of regulatory reviews:** A regulatory body undertakes a review against its own requirements and seeks to leverage existing reviews made by other regulatory bodies. Further information about leveraging reviews can be found in Ref. [2].

A combination of these three cooperation types is possible as described in Ref. [2].

Information about the reactor design will typically originate from the vendor and be presented to and reviewed by regulatory bodies or associated TSOs from one or more States. Some of this information may be controlled information (see Section 3). It is acknowledged that the controlled information may also belong to other interested parties involved in the design development and future operation, such as engineering design companies contracted by vendors or applicants. For simplification, this publication refers to the design information owner (DIO) as follows:

- The DIO is the originator, designer, or owner of the controlled information associated with the systems or components of a nuclear power plant (NPP). This can be one single entity (e.g. the primary reactor vendor) or several entities. In licensing submissions, the DIO may also be the applicant aiming to become a reactor operating organization using a specific vendor's design as part of a commercial agreement. For the purpose of this publication, the DIO is viewed as a single entity encompassing all owners of controlled information associated with a reactor design.

It can be anticipated that a prerequisite for the exchange of information is a commitment to commercial confidentiality from the data recipients. In the case of a single State, this will likely be covered by domestic law but in a cooperation involving regulatory bodies from different States, this will be formalized through information sharing agreements (e.g. confidentiality and non-disclosure agreements (NDAs)) with individual DIOs. Section 2.2 discusses the types of information needed for the reviews. For the sharing of information among regulatory bodies from different States, the overarching information sharing agreements may also include details of the processes to be followed. A more extended description of this type of information is provided in Section 2.3.

The documents that support the reviews, regardless of the number of regulatory bodies involved, can be grouped into two general categories:

- Documents submitted to the regulatory bodies by the DIO.

- Documents generated by regulatory bodies as a consequence of the review. The results of the review process will naturally contain information about the design along with the review results and conclusions drawn by each regulatory body.

2.1. NATIONAL LICENSING REVIEW PROCESSES AND PRE-LICENSING REGULATORY ENGAGEMENTS

There are differences in the ways that States license reactor designs. These include the regulatory approach used, the time needed to process the licence, and the different licensing stages involved. A general overview of the application of a graded approach in regulating nuclear installations is provided in Ref. [5]. This methodology could be considered by States for use in reviews prior to the licensing process (i.e. pre-licensing reviews) or during the licensing process itself.

Several States have well-developed regulatory practices, policies and expectations for engaging with potential applicants and subsequent reviews of formal licence applications. Some of these States require regulatory bodies to be transparent and to give guidance on the regulatory framework and processes, while only providing formal feedback or decisions in specified circumstances. In some States, however, even though not explicitly defined in the national standards and/or regulations, there are also established policies and practices for early exchange of information between a reactor developer and the regulatory body. While related information is provided in Annex II, the following are examples of such cases:

- The United States Nuclear Regulatory Commission (US NRC) has a policy of encouraging early discussions with potential applicants, such as utilities and reactor designers, prior to submission of a licence application. The purpose of these discussions is to offer licensing guidance, become familiar with new concepts or novel design features, allow for early identification and resolution of technical and policy issues that could affect licensing, and identify gaps in information and data necessary to establish the safety of the design [6], [7].

- The Argentine Nuclear Regulatory Authority holds meetings with interested parties in advance of licensing to provide guidance on the national licensing system. Then, in the most advanced stage, and closest to the beginning of the licensing process where the decisions about the investment and project times begin, a memorandum of understanding (MoU) is signed. This gives the DIO firm guarantees of the applicable regulatory framework that will determine the conditions of the licence.

In addition, there are a few States with a formal review process prior to a licence application. While related information is provided in Annex II, the following are examples of such cases:

- By issuing a design certification, US NRC approves an NPP design, independent of a specific site or an application to construct or operate a plant, by addressing various safety and security issues during the application review. During this process, US NRC notifies all interested parties (including the public) about how and when they can participate in the regulatory process, which may include participation in public meetings and rulemaking activities related to design certifications.

- The Canadian Nuclear Safety Commission (CNSC) provides a pre-licensing review process for assessing a vendor's reactor design [8]. The purpose of this vendor design review is to evaluate a supplier's NPP design. The vendor design review considers the design areas of the reactor: safety, security and safeguards.

- The United Kingdom Office for Nuclear Regulation (ONR) and the Environment Agency (or when appropriate Natural Resources Wales), conduct a generic design assessment [9], [10]. The objective of the generic design assessment is to provide confidence that the proposed design is capable of being constructed, operated and decommissioned in accordance with the national standards of safety, security, environmental protection and waste management. The generic design assessment is not a mandatory process, but because of its inherent benefits, it is expected that it will usually be requested for new NPPs intended for construction.

- The French Authority for Nuclear Safety and Radiation Protection (ASNR) carries out a review of safety options. The purpose is to assess whether or not the safety options comply with French regulations. It has the following characteristics: it is optional, at the applicant's choice; it is related to a generic and conceptual design; it ends with an official and public statement by ASNR about the safety options of the project; with a list of expected changes that will be needed before a construction licence can be issued.

The Republic of Korea (ROK), Finland and Sweden are at various stages in planning or developing a process to support pre-licensing engagements with vendors. The ROK Nuclear Safety and Security Commission (NSSC) is planning to develop the innovative-SMR regulatory preparation group whose members are regulatory bodies, developers and expert groups. This group plans to launch a pre-application review service prior to a licence application, and identify regulatory issues using a technical gap analysis, review measures for safety, and prepare regulatory positions for any issues that arise. The results of this service would not be legally binding. In Sweden and Finland, there are ongoing discussions and suggestions to establish pre-licensing review processes. The Swedish Radiation Safety Authority (SSM) has proposed to the government that SSM be given the mandate to perform a non-binding review that is available for different reactors and flexible with regard to the issue or question to be reviewed (i.e. a specific design principle, a greater scope, or a conceptual review). In Finland, there is a proposal to establish a pre-licensing review of reactor designs to be performed by the Radiation and Nuclear Safety Authority (STUK). Different options are under discussion; for example, the review could be optional or mandatory, could include both design and vendor/applicant information and could have an associated validity period. In both Sweden and Finland, screening criteria to enter a pre-licensing review are being discussed. Collaboration among regulatory bodies can take place at the different stages of a regulatory review presented above.

2.2. INFORMATION INPUTS TO REVIEW

Regulatory bodies may follow a process to share information when cooperating in regulatory reviews, Figure 1 provides an example of the process that may be adopted. The process begins with the DIO submitting a formal request for review together with the supporting documents. These will vary in accordance with the nature of the review and the requirements of the regulatory body but will usually include a safety analysis report (SAR) and other documents assessing security and safeguards.

During the course of a review, the regulatory bodies may submit to the DIO questions to obtain additional information that is needed to make a regulatory decision or complete the review. In some States, the formal request is called a request for information (RFI) or request for additional information (RAI). The information requested typically addresses missing, incomplete, inconsistent, or unclear information within the application or associated documents. Questions for additional information can also be used by regulatory bodies to gain access to inputs to DIO's calculations in order that the regulatory bodies can perform independent calculations, as necessary, to support the review. Meetings at a DIO's premises can be used to gather data, to ensure a common understanding of a request, to provide clarifying information, and to discuss the responses. In the course of these communications between the regulatory bodies and the DIO, the RFIs or RAIs can be eliminated, revised, or supplemented prior to being formally sent to the DIO.

The DIO will then provide the regulatory bodies with a document responding to the requests and questions. The response could also include proposed mark-ups that revise and supplement the information in the application or the submitted reports to support the review, findings and/or feedback.

The regulatory bodies can document the review, findings and/or feedback in documents such as safety evaluation reports (SERs), technical reports, summary reports, or assessment reports. The regulatory bodies can also document the initial, interim or draft review, including the remaining issues that need to be resolved, in documents such as draft SERs with open items or Interim Design Acceptance Confirmations (iDACs).

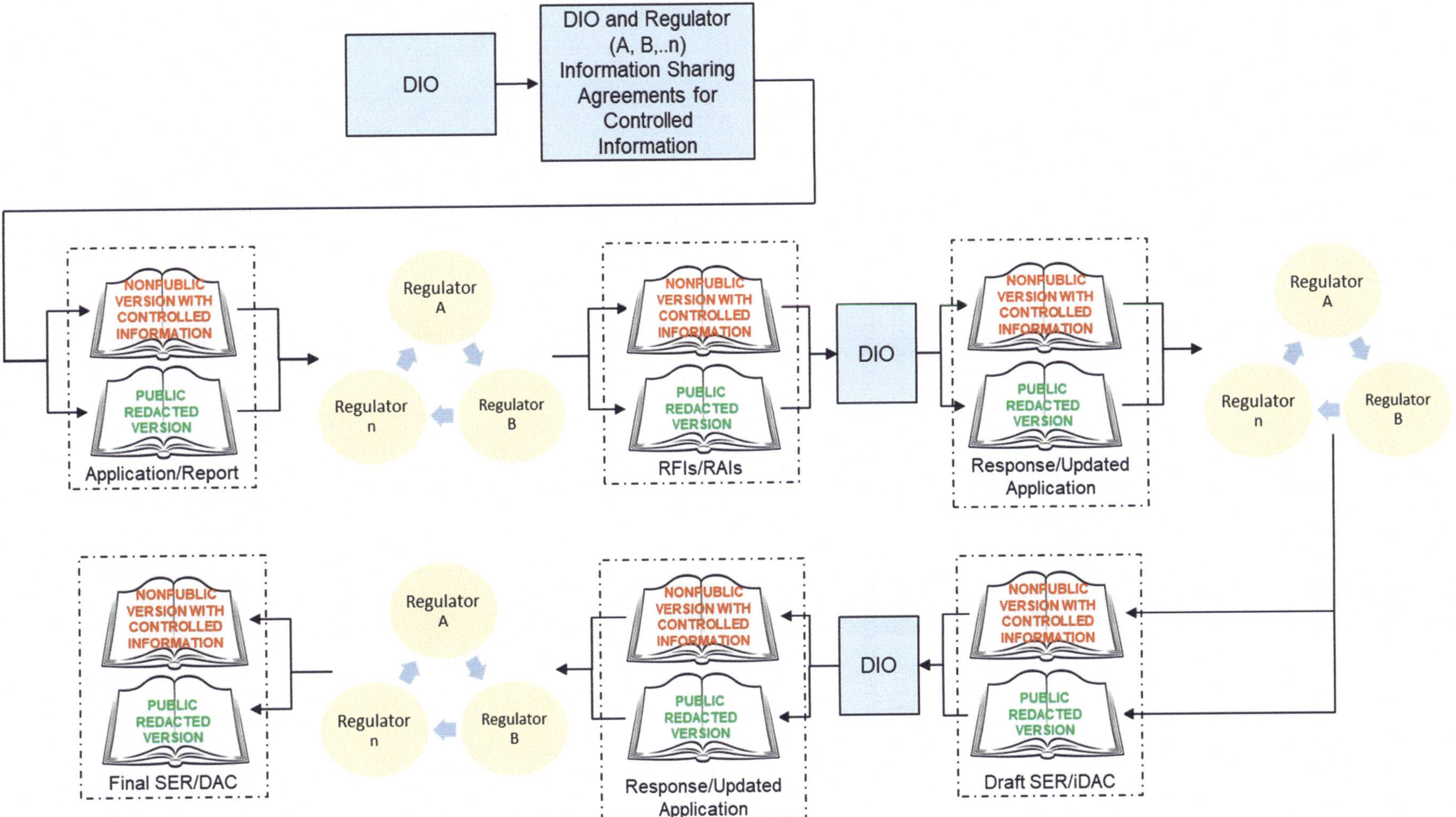

FIG. 1 Typical Information Flow and Process Envisioned for Supporting Cooperation in Regulatory Reviews with One or More Regulatory Bodies.

Note: If there is more than one regulatory body conducting the review, once information sharing agreements covering controlled information are signed with the DIO, regulatory bodies may communicate continually over the review cycle.

2.2.1. Types of information needed for regulatory review

The different topics listed in this subsection are typically part of regulatory reviews of advanced reactors. These topics may also be considered by regulatory bodies when leveraging reviews performed by other regulatory bodies or when participating in joint reviews and collaborative reviews.

Regulatory reviews may be requested when a reactor is in the basic design phase, where the focus of the review is at the system level. Nevertheless, for such a review to be effective, the information needed is wide ranging. For example, the information may include the following:

- Applicant organization: management system for safety, quality, supply chain, etc.;

- Design information: description of the systems, safety objective and acceptance criteria, design criteria and design approach;

- Safety information: consideration of design basis accidents and design extension conditions, the independence of safety systems, defence in depth levels, practical elimination of conditions that would lead to an early or large radioactive release, safety margins to avoid cliff edge effects, the capabilities for heat transfer to an ultimate heat sink, and the use of non-permanent equipment for accident management;

- Civil engineering works and structures;

- Bounding site characteristics (e.g. susceptibility to internal and external hazards, generic site envelope);

- General lifetime considerations: construction, commissioning, operation and decommissioning requirements of the facility;

- Radiation protection measures;

- Radioactive waste management;

- Safety analysis: deterministic safety analysis (DSA) and probabilistic safety assessment (PSA);

- Human factors engineering;

- Integration of safety, security and safeguards in the design (3S Concept);

- Design features for decommissioning.

Additional examples of information needed by regulatory bodies for reviews as part of pre-licensing or licensing processes are provided in Annex II.

The DIO will have performed many calculations and analyses to justify the safety of the proposed design. As part of the submissions, the DIO also describes the computer models, assumptions, parameter inputs and results of these calculations and analyses. These details include modelling and simulation tools, software and/or computational analyses supporting the safety analysis of the design in areas such as:

- Reactor physics, including criticality;

- Fuel characteristics and performance;

- Thermohydraulic behaviour;

- Severe accidents;

- Source terms;

- Offsite consequence analysis;

- Radiation protection and health physics;

- Materials and component integrity;

- PSA.

During reviews, the regulatory body will assess whether the DIO's calculations and analyses are complete, demonstrated to be sufficiently verified and validated, meet quality requirements, and have been undertaken by suitably qualified experts. The approach to the review of technical submissions could be as follows:

- Review the description of the computer code models, related assumptions, input parameters and outputs as presented in the DIO's application and report;

- Review the implementation of the DIO's analysis and modelling tools (e.g. a computer code, and associated computational analysis);

- Obtain a copy of the DIO's input datasets for computer models and perform confirmatory analyses and sensitivity studies on the DIO's software;

- Obtain detailed design information from the DIO and develop and run the independent computer model of the design using computer codes used by the regulatory body.

Confirmatory analyses can also be used to focus the review of the regulatory bodies, lessen the need for RFIs/RAIs, and decide whether an information gathering meeting at a DIO's premises would be helpful. In addition, it is anticipated that modelling and simulation tools, computer codes, computational analyses and input dataset will likely contain controlled information, as discussed in Section 2.2.2.

Recommendations on additional inputs to a review are provided in IAEA Safety Standards Series No. SSG-61, Format and Content of the Safety Analysis Report for Nuclear Power Plants [11]. Paragraphs 2.5–2.10 of SSG-61 [11] provide recommendations on the information to be provided in a licence application at different stages of the project, and provide examples of the level of detail needed. Paragraphs 2.12–2.14 of SSG-61 [11] provide recommendations on a possible structure for a safety analysis report.

Other topics to be addressed at various stages of reviews may include security, safeguards, environmental protection, financial guarantees, liability for nuclear damage, decommissioning funding, and consultation with interested parties. Information related to security may include:

- Details of physical protection systems and any other security measures in place for nuclear and other radioactive material, associated facilities and activities, including information on guard and response forces;

- Information relating to the quantity and form of nuclear and other radioactive material in use or storage, including nuclear material accounting information;

- Information relating to the quantity and form of nuclear and other radioactive material in transport;

- Details of computer systems, including communication systems, that process, handle, store or transmit information that is directly or indirectly important to safety and security;

- Contingency and response plans for nuclear security events;

- Information relating to employees, vendors and contractors (uses, identities, capabilities etc.);

- Threat assessments and security alerting information;

- Details of sensitive technology;

- Details of vulnerabilities or weaknesses that relate to the above topics;

- Historical information on any of the above topics.

Some States include considerations for IAEA safeguards in the early stages of the review of reactor designs, as this presents a good opportunity for making design modifications that reduce the burden of safeguards implementation, and potentially increase its effectiveness, during operation. The consideration of safeguards at an early stage of design development is an industry best practice known as safeguards by design (SBD). It occurs earlier than the mandated provision of design information under a comprehensive safeguards agreement (CSA) with the IAEA, which is triggered by a State's authorization to construct. Since high level safeguards obligations do not differ significantly from one CSA State to another, a significant degree of harmonized industry preparation for international safeguards requirements can be achieved through this process. This can potentially reduce the cost and time (and resulting risk to schedule) needed to incorporate these safeguards. The type of facility-specific information shared with the regulatory body and IAEA in the SBD process includes the facility layout, as well as the location, flow and nature of nuclear material. Information exchange with other regulatory bodies relevant to this level of SBD (for similar reactor design types) would not include controlled, State-specific, detailed design information, nor facility-specific details about future safeguards measures as these are determined at a later stage of safeguards implementation with the IAEA. Collaboration among regulatory bodies in this area would help to achieve a generic safeguards approach for the design type under examination and would enhance industry awareness of high level safeguards implications generally.

2.2.2. Information gathering at a DIO's premises

Information gathering at a DIO's premises, with access to either physical or digital information, is an optional documented activity where the regulatory bodies, or another entity on their behalf, evaluates and discusses the adequacy of objective evidence that supports the safety analysis described in the DIO's submission. Since the objective evidence can include reports, calculations and analyses owned by a DIO and can contain controlled information, information gathering at a DIO's premises is typically non-public.

Information gathering at a DIO's premises is a tool that could be used to improve the efficiency and effectiveness of regulatory bodies performing future joint and collaborative reviews by helping to efficiently gain understanding, verify information, and/or identify additional information that will be needed to support regulatory decision making. The information gathered may include topical reports, design calculations (inputs and methodology, including internally developed codes), design drawings and white papers[1]. Access to these documents may allow regulatory bodies to:

---

[1] A white paper is a document that aims to concisely describe a topic associated with an advanced reactor, including its major structures and components and their layout, mode of operation, cooling systems, safety systems and the general approach to safety, manufacture, siting and construction.

- Gain a better understanding of the detailed information contained within the submission and, more generally, confirm the regulatory bodies' understanding of the proposed design.

- Identify additional information necessary for a regulatory decision which the DIO will then provide as a supplement to the review submission.

- Establish a better understanding of potential areas of concern and issue clear requests for information. This would help the DIO to provide timely responses of sufficient quality. Efficiencies can also be obtained if the DIO volunteers to supplement its review submission to avoid the need for regulatory bodies to issue formal requests for information.

2.3. INFORMATION RELATED TO THE REVIEW OUTCOMES THAT IS NEEDED FOR COOPERATION BEETWEEN REGULATORY BODIES

2.3.1. Collaborative and joint reviews

Section 2.2 discusses the typical topics and information reviewed by regulatory bodies during pre-licensing or licensing. This also applies to collaborative or joint reviews.

2.3.2. Regulatory bodies leveraging reviews of other regulatory bodies

This subsection defines specific information, which in the course of cooperation needs to be shared between a regulatory body carrying out a review (e.g. Regulator A) and a regulatory body or government, who wishes to have access to the review (e.g. Regulator B), with the likely involvement of the DIO.

Figure 2 shows examples of how Regulator B could leverage reviews of new reactor designs already completed by Regulator A. Regulator A may also share with Regulator B other documents such as RFIs, SERs, technical reports, summary reports, assessment reports, etc. Regulator A may also generate new documents, such as presentations and documents answering questions from the other regulatory bodies, to help the other regulatory bodies better understand Regulator A's licensing process and regulations, country-specific design differences and the resolution of technical issues identified during the review. Detailed information on approaches to leverage regulatory reviews can be found in Ref. [2].

In the course of implementing the interaction between Regulator A and Regulator B, in the review of design documents, the information below may be needed:

- Information about the outcome of the review;

- Information about review criteria;

- Information in relation to the experience of Regulator A;

- Information about future interactions between Regulator A and Regulator B.

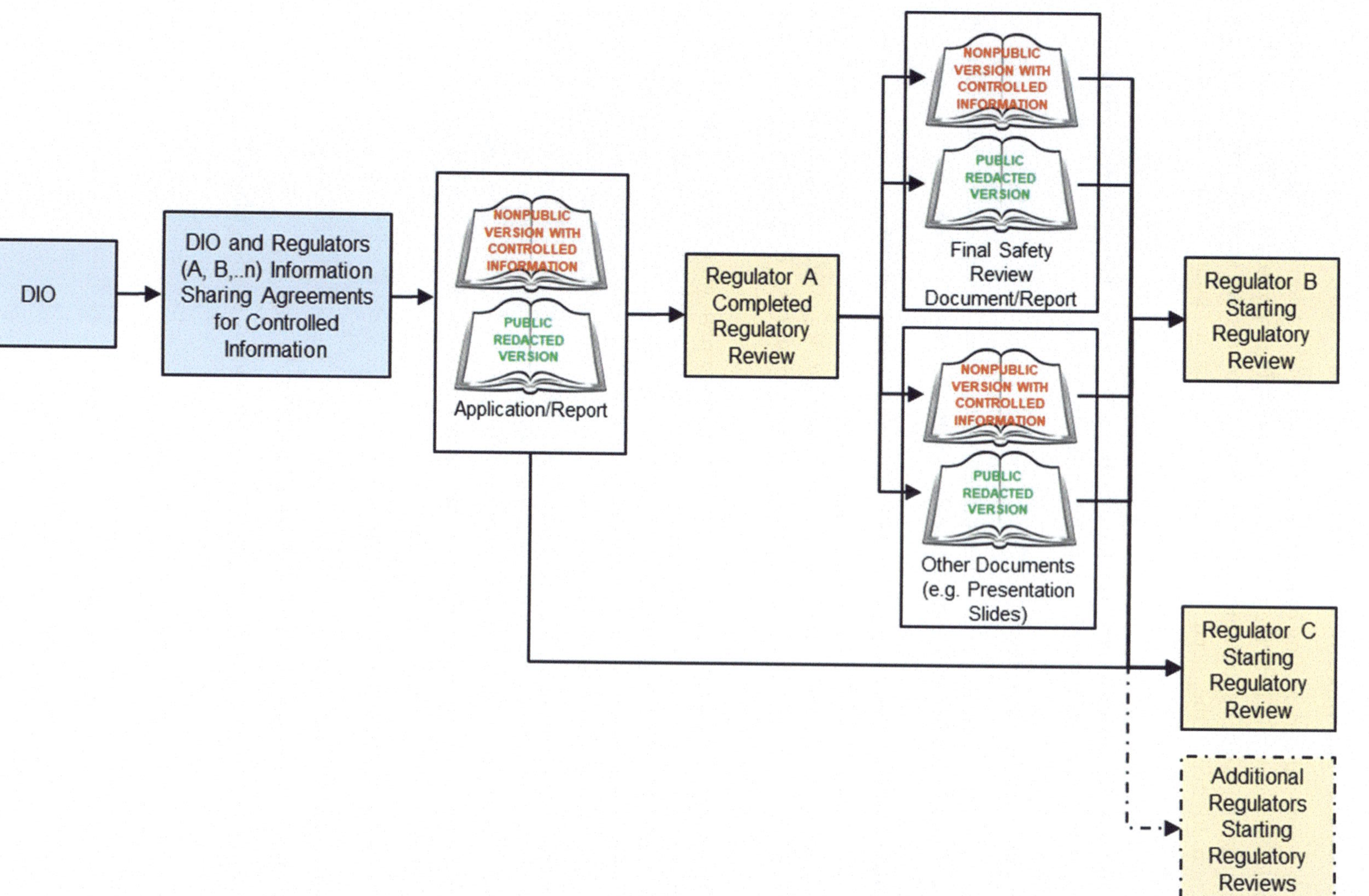

FIG. 2 Leveraging completed regulatory reviews (for readability, additional minor steps that may occur in the process are not included (e.g. RFIs between entities)).

2.3.2.1. Information about the outcome of the review

The regulatory bodies will document the outcome of the regulatory review in SERs, technical reports, summary reports, assessment reports, or other documents. The regulatory bodies may also present the results as an initial, interim or draft review that describes the remaining issues that need to be resolved, such as a draft SER with open items or iDACs. As a result of the review carried out by Regulator A, the following list of documents may be transferred to Regulator B:

- Expert review carried out by Regulator A;

- A list of DIO documents assessed in the course of the review;

- A list of regulatory documents and criteria used for the review; regulatory documents can have either international or regional status, or documents applicable in the country of Regulator A and/or Regulator B can be used;

- Information, such as meeting minutes, about interaction with DIO in the course of review;

- Results of alternative calculations made by Regulator A and general information about assumptions and approximations used, if applicable;

- For advanced reactors for which the design reference has changed (where the completed Regulator A's review was based on a design reference and does not correspond to the design under review by Regulator B), the DIO may perform an assessment of the changes and modifications and provide these to Regulator B for review.

2.3.2.2. Information about review criteria

Information about the regulatory documents and the criteria used in the regulatory review may be provided by Regulator A as part of the set of documents provided to Regulator B with the results of the review. If necessary, upon request, Regulator B could be provided with additional information such as controlled regulatory documents, tools and criteria used for the review, as well as a justification of the applicability of the review criteria.

2.3.2.3. Information in relation to Regulator A experience

The relevant experience of Regulator A may be provided to Regulator B for confirmation. This information is expected to be non-controlled and may include the following:

- Background information about the role and competence of Regulator A in the country, based on the recommendations provided in IAEA Safety Standards Series No GSG-13, Functions and Processes of the Regulatory Body for Safety [12];

- Reports with information about the IAEA Integrated Regulatory Review Service missions carried out in the country;

- Information about the national regulatory framework and requirements for the type of installations considered;

- Information about experience gained through reviews of different nuclear installations, including NPPs, research reactors and innovative facilities.

2.3.2.4. Information about future interactions between Regulator A and Regulator B

Confirmation of future interactions is especially appropriate for embarking countries that are lacking in experience of regulating the safety of nuclear installations. For example, assistance may be needed in the development of the regulatory framework, the safety supervision at nuclear installations, and in the preparation and carrying out of the inspections. On this issue, Regulator A and Regulator B may determine the format of future interactions and might even develop a long term cooperation programme provided that all parties are in agreement. On the request of Regulator B, Regulator A can additionally arrange the following:

- Meetings between Regulator A and Regulator B, including meetings with experts before Regulator B starts the review;

- Lectures on regulatory approaches of Regulator A, with the possibility of adjusting the topics of lectures depending on the request of Regulator B;

- Consultations provided by Regulator A to Regulator B, including the issues concerning the results of the review carried out by Regulator A.

2.4. INFORMATION FOR THE PUBLIC

Many States share safety relevant information with the public to ensure openness and transparency about the regulatory reviews of advanced reactors, such as the data in a SAR, SER, or other related documents. In many cases this will involve translation into different languages by the DIO of, at least, key documents. The sharing of information can either be done via the DIO's website or the regulatory body's website. Similarly, the review findings are typically widely published and include ancillary information such as RFIs, technical reports, summary reports and assessment reports as well as accounts of interactions between participating regulatory bodies. Note that country-specific regulations will also apply but, in general, published data typically contains conclusions and results without necessarily all the associated details with supporting technical analyses.

When DIO information is made public, it is the responsibility of the author of the documents, in consultation with the DIO, to ensure that it does not contain any controlled information; this includes trade secrets and information that could hinder security (see para. 2.29 of SSG-61 [11]). As shown in Fig. 1, if the documents that are submitted to a regulatory body by the DIO or the documents generated by the regulatory body contain controlled information, then the documents need to be marked and handled appropriately, and a separate redacted version needs to be made available to the public. This also applies to relevant materials covered by export controls and cases where the DIO may require an export approval to share information with a foreign regulatory body.

The international agreements related to environmental protection and environmental impact assessment (e.g. the Aarhus Convention and the Espoo Convention) impose a duty to disclose information to the public that is relevant to environmental decision making, and they stipulate the right of the public to access information, to participate in decision making and to have access to justice in environmental matters. While these treaties do allow sensitive and controlled information to be withheld, this is not unconditional.

2.5. INFORMATION REQUIRED FOR THE DEVELOPMENT OF SAFETY AND SECURITY INFRASTRUCTURE

The main scope of this subsection is to identify additional design phase specific information that may be important to support regulatory review and assessment activities regarding advanced reactors in embarking countries. This is of particular importance for the consideration of SMRs that present areas of novelty that might not be reflected in international requirements and recommendations such as in the IAEA safety standards [13]. Embarking countries in particular can therefore benefit from learning about how regulatory bodies have considered SMR-specific issues in their regulatory frameworks and reviews. This subsection includes siting and site evaluation studies as they pose challenges to embarking countries when considering comprehensive licensing reviews by other regulatory bodies.

The information identified in Section 2.2 is intended to be sufficient for States that have already licensed an NPP. Additional information may be needed, however, by others such as embarking countries that are currently licensing an initial unit and States that have previously only licensed a nuclear fuel cycle facility or a research reactor to help the development of the safety, security and regulatory infrastructure. These additional information and support needs will be both general and technology-specific; the latter could be particularly important if the reactor type under consideration is relatively new (e.g. an SMR) for which there is more limited information on establishing adequate regulatory frameworks. This subsection focuses on States that have decided to embark on a nuclear programme and whose first nuclear installation is to be an SMR. These States need to develop their safety and security infrastructure and will benefit from support from an experienced regulatory authority.

2.5.1. Additional information designated as safety and security related

Establishing the regulation and authorization system of an embarking country involves establishing a framework that includes core regulatory activities, such as licensing and authorization, and the promotion of concepts such as safety culture, defence in depth, safety classification and safety assessment. The framework is to be guided by the ten fundamental safety principles in SF-1 [4], focusing on comprehensive safety measures and a commitment to continuous assessment and improvement.

For embarking countries interested in SMRs, it is important also to define and understand terms that are SMR-specific or important for SMRs, such as first of a kind (FOAK), multi-module unit, modular reactor and proven technology. During the development stage, nuclear professionals in embarking countries will need to familiarize themselves, not only with the technology and important terms and concepts such as those listed above, but also with the general approach to regulation, as expressed in IAEA safety standards and in the regulations and guidance developed by States with long-standing nuclear programmes. In order to establish a safety–security infrastructure, embarking countries may use IAEA publications, training tools and review services. Embarking countries may also benefit from the IAEA's training, self-assessment modules and peer review services for their infrastructure development.

The embarking countries may benefit from learning about the regulatory reviews already performed on SMR designs to enhance their understanding of SMR specific issues, including:

- Information about regulations in States that have already reviewed the SMR of interest, including detailed procedures and review guides such as standard review plans.

- If available, modified or specific requirements, procedures, criteria and alternative approaches for SMRs and FOAK features from experienced States.

- The importance of site-specific issues such as specific natural and human-made hazards, the associated SMR design response, and the arrangement for the emergency planning zone.

2.5.2. Additional information related to site-specific issues

Embarking counties need to understand the importance of site-specific issues when they are trying to learn from regulatory reviews already performed. Embarking countries may benefit from the regulatory reviews performed by experienced States with similar site characteristics. If however, the previously reviewed SMR has different site characteristics from the actual site, this may require the actual plant design to differ from the previously reviewed case.

Embarking countries may wish to work with experienced States to understand how other regulatory bodies have considered site-specific issues and the potential impacts on the design. Embarking countries may benefit from additional site-specific documentation produced by experienced States, to understand how the suitability of the site was evaluated. This could include information on site specific hazards, emergency preparedness and response, and site evaluation studies.

2.5.3. Additional information related to nuclear safety computer codes

If embarking countries are to reach a high level of readiness and confidence in the use of safety analysis computer codes, they may need access to background information on the proper application of these tools to DSA and PSA. In addition, knowledge of applicable industrial standards and industry-specific requirements for verification, validation and testing are necessary for an effective regulatory environment. This effort may need to utilize commercial entities for specialized services such as confirmatory analyses, analysis validation, or other similar technical consulting engagements.

Embarking countries will also need to have the necessary competences in using computer codes. Identifying proper conditions for a regulatory body to carry out separate calculations with different codes is an area that needs to be explored. Even if the regulatory body will only repeat the DIO's calculation using the same computer code, it is necessary to know the steps. If the regulatory body decides to do a separate calculation, it is necessary to learn the verification process for the code that will be used, as well as where to find the necessary input data. In addition, on the issue of a testing facility for a FOAK technology, embarking countries will need to learn how the regulatory body might build the necessary confidence and competence to accept a proposed test facility along with the appropriate test specifications and scaling of the output data to the necessary in-service conditions. This may be a specific topic for engagement between embarking countries and experienced States.

3. IDENTIFICATION OF POTENTIAL IMPEDIMENTS TO SHARING INFORMATION

This section discusses potential impediments that regulatory bodies might face when sharing information identified in Section 2. Principally, it identifies the categories of information that regulatory bodies might not be able to readily share with one another. This is referred to in this section as controlled information and it includes information designated as (a) proprietary (e.g. trade secrets, confidential commercial, intellectual property and some types of financial information); (b) nuclear security related (NSR), including information designated as classified or relating to safeguards; (c) import/export controlled; and (d) pre-decisional information. Where sharing is permitted it is expected that it would be subject to a written agreement establishing appropriate controls.

These written agreements are discussed in more detail in Section 4.3; however, there are common impediments that States may face, regardless of the characteristics of the information:

- **Need to know.** Sharing of controlled information generally involves a threshold determination that the proposed recipient of the information has a requisite need to know, or a lawful purpose to receive the information (i.e. that the proposed recipient needs the information to perform a valid recognized task and that the type and amount of information to be shared is appropriate and commensurate for the task to be performed). For bilateral exchanges of controlled information where two States are performing a parallel review of the same design, establishing this need to know will generally not be difficult. However, in a multilateral setting it may be significantly more difficult for a State to determine that all parties to that agreement have the same requisite need to know, especially if all parties to the multilateral exchange are not actively engaged in review at the same time.

- **Differences in domestic legal frameworks.** Each State will have its own domestic laws, regulations and policies for the handling and protection of controlled information when in its own possession or control. This can include a number of discrete subject matters, such as:

  - Requirements for the appropriate marking of controlled information when in a regulatory body's possession (e.g. use of prominent banner markings or watermarks; portion marking; unique labels and abbreviations);

  - Requirements for establishing appropriate controlled environments when controlled information is being actively used (e.g. only permitting the use or discussion of controlled information in environments that minimize or eliminate the likelihood of unauthorized disclosure);

  - Requirements for the appropriate handling and storage of controlled information in physical format (e.g. requirements for the controlled information to be locked and inaccessible to unauthorized persons when not actively in use);

  - Minimum cybersecurity requirements for information systems that access or store controlled information in digital format, including not only computer networks and drives but also other technologies such as networked printers, copiers and fax machines that temporarily store controlled information;

  - Requirements for the appropriate destruction of controlled information to ensure it is not recoverable;

  - Requirements for additional bilateral mechanisms or controls to be placed on certain export-controlled information (e.g. use of bilateral, treaty-level agreements outlining non-proliferation obligations; or use of diplomatic government-to-government exchanges to provide peaceful use and non-proliferation assurances).

Disparate domestic legal frameworks for the protection of controlled information might create impediments for States when negotiating written agreements establishing the appropriate protection of controlled information to be shared with one another. A State might also lack overall confidence that the information it shares will be appropriately protected if the standards between the two are significantly different, or if one State lacks experience in protecting certain kinds of information, for example:

- **Ability of recipient to guarantee non-disclosure.** It is assumed that all States have structures to protect controlled information. Many States have legislation related to public transparency that makes records created by or in the possession of governmental entities publicly available by default, unless a recognized exception to the law permits non-disclosure. Depending on the nature of the information shared or the breadth of recognized exceptions under domestic law, some States may be unable to provide complete legal assurances in a written agreement that information in their possession will not be disclosed publicly. Instead, these States may only be able to guarantee that the information will be withheld from the public to the fullest extent allowed by its own domestic law.

- **Translation.** For cases where the documents being shared were originally prepared in a different language, the documents could need to be translated into the chosen language of the other regulatory bodies participating in cooperation such as a joint or collaborative review to facilitate transparency and a better understanding of the information. It is important that the translated documents continue to maintain appropriate markings and redactions of any controlled information. In addition to written documents, issues concerning translation can also arise in the context of verbal discussions, particularly in situations where regulatory bodies are collaborating to produce a joint document.

All of the impediments identified above are generally applicable and would be present regardless of the category of controlled information. Additional impediments that are specific to particular types of controlled information are identified in the following subsections.

3.1. IMPEDIMENTS WITH MANAGING AND SHARING CONTROLLED INFORMATION

3.1.1. Proprietary information

This subsection identifies potential impediments that regulatory bodies could encounter when managing and sharing proprietary information, or information designated as trade secret, confidential commercial, financial, or intellectual property. A trade secret is a valuable piece of information that could damage the company if revealed, so that it will need to be treated as confidential. For example: a commercially valuable plan, formula, process, or device that is used for making, preparing, compounding, or processing trade commodities and that can be said to be the end product of either innovation or substantial effort.

Confidential commercial, financial, or intellectual property is valuable data or information which is used in business and is of a type customarily held in strict confidence or regarded as privileged and not disclosed to any member of the public by the person to whom it belongs. Information may be protected under intellectual property law and may undergo changes in ownership, depending on the commercial arrangements of licensed entities owning and operating an NPP. Inclusion of intellectual property information may mean that the rights owner has to protect their asset from undue use or duplication. Hence, allowing the recipient of information (i.e. a regulatory body) to share certain information may be hindered or forbidden as this might jeopardize the protection of that information under intellectual property laws.

Consequently, the following outlines the associated impediment discussion. Regulatory bodies may be obliged to obtain the DIO's permission before disclosing the DIO's information to third parties, granted that they would in turn protect the information. These obligations may derive from domestic legislation prohibiting government officials from disclosing business trade secrets or other confidential information received from another person without authorization. Additionally, regulation or policy may require regulatory bodies to provide businesses with assurances that confidential information submitted to the regulatory body will not be released without the DIO's consent. In the case of existing intellectual property rights, sharing may be contingent on the approval of the owner of the controlled information.

3.1.2. Information designated as nuclear security related

This subsection describes information that a regulatory body has designated as needing specific control or protection because of concerns that disclosure may compromise nuclear security or contribute to a nuclear security event. This information is generally referred to in this publication as NSR information.

The overall objective of a State's nuclear security regime is to protect persons, property, society and the environment from harmful consequences of a nuclear security event. Groups or individuals wishing to plan or commit any malicious act involving nuclear or other radioactive material or associated facilities may benefit from access to NSR information. It is expected that all States will have domestic laws, regulations and policies that govern the identification, classification, management and confidentiality of NSR information. Examples of NSR information can be found in Annex III of IAEA Nuclear Security Series No. 23-G [14]. Generally speaking, this information may include details of physical protection systems and measures; details of sensitive nuclear technologies; threat assessments and design basis threats; details of vulnerabilities or weaknesses associated with items; or any other information that a regulatory body deems could be helpful or provide assistance to a malevolent actor. With respect to information pertaining to an advanced reactor design, this may include information referred to as security by design or safeguards by design (e.g. details of how certain elements of a design are intended to guard against a design basis threat, or details of how a design is intended to facilitate safeguards inspections or detection of diversion of nuclear material).

NSR information exists on a spectrum that will generally determine the level of control applied to that information and the ability or willingness of a State to share it. Protection of NSR information within this spectrum may be mandated by legislative act or regulation, or may be governed by non-legislative policies, depending on the State, for example:

- On the high end of this spectrum, it is generally expected that the most significant NSR information would be classified under a State's legal framework (e.g. information designated by the State as 'secret,' 'confidential,' 'restricted,' or another classification). This would include information whose unauthorized disclosure could compromise national security interests. It is assumed that a regulatory body would be unable to share such information, without approval from the highest levels of government which is granted in extraordinary circumstances.

- NSR information that does not rise to the level of classified data but that is still sufficiently detailed to present significant security concerns if disclosed will also be closely held by a regulatory body under its domestic laws and policies. This intermediate level of NSR information would not readily be shareable, without high level approval within the regulatory body or within the government, depending on the State.

- A regulatory body may possess or generate unclassified information that does not rise to the significance levels described above, but that the regulatory body chooses to withhold from public view because of concerns that the information may assist a malevolent actor. Regulatory bodies may have more discretion to share this type of information, depending on the level of detail, provided that such sharing is done pursuant to a written agreement that ensures the recipient will respect its confidentiality and protect it accordingly while in its possession.

Consequently, the following outlines the associated impediment discussion. A regulatory body's willingness or ability to share NSR information with another regulatory body will generally depend on the regulatory body's subjective judgment as to the severity of the harm that could occur if the information is released or compromised. In some cases, domestic law or policy will severely restrict or prohibit the sharing of detailed information of a security nature, such as information relating to a State's design basis threat or detailed security measures for the physical protection of nuclear material or equipment vital to safety. Sharing of such information may also necessitate the approval of senior government officials. In other cases, information may be NSR but at a level of detail where a regulatory body is willing to share the information pursuant to a written agreement that provides for its protection.

3.1.3. Information designated as export controlled

Individual States may have additional requirements that apply to the exchange of non-public, proliferation-sensitive nuclear technologies. This is generally referred to as 'export controlled' information. For purposes of this publication, export-controlled information may include technology, technical data, software, or technical assistance that could contribute to an unsafeguarded nuclear fuel cycle or nuclear explosive device, including technology associated with advanced reactor designs. In addition, an export of technology may take the form of tangible information exchanges (such as through physical documents), as well as intangible transfers (such as through verbal discussions, electronic data exchanges, or remote access to information which is electronically stored in another country). Export control in this context does not refer to actual exports of nuclear material or nuclear equipment and components. Depending on the State and the information transfers taking place, these export approvals may rely on or require the use of additional bilateral mechanisms or controls, such as in the form of treaty-level cooperation agreements with non-proliferation provisions, or government-to-government diplomatic peaceful use assurance exchanges.

If information is subject to export controls, then typically a licence or other approval from the relevant authority within government is needed before the information can be exported or provided to another State or to a citizen of another State. Depending on the State, these licences or approvals may be granted on a fact-specific basis or generally preauthorized (e.g. an open licence or a general licence), depending on the nature of the information and the destination of the export. Regulatory bodies may also have requirements for additional bilateral mechanisms or controls to be placed on certain export-controlled information (e.g. use of bilateral, treaty-level agreements outlining non-proliferation obligations; or use of government-to-government assurances to provide peaceful use and non-proliferation assurances).

Export controls may introduce a high degree of complexity in the sharing of sensitive information, especially given the variability in legislative frameworks, policies and protocols among States. Each State's specific export regulations dictate not only the types of controlled information that can be shared but also impose specific conditions on export licences. These licences often contain stringent conditions, including end user undertakings that restrict further re-exporting, thereby requiring recipients to agree not to disseminate controlled documentation without explicit permissions. Regulatory bodies and their TSOs that operate across national borders may face additional challenges. For example, organizations may need separate licences to re-export information back to the original licensors or to other regulatory bodies. This complexity is heightened in situations where export-controlled discussions take place verbally, demanding a proactive approach to comply with export licensing conditions. The process of obtaining export licences itself can be lengthy, often involving multiple interested parties and governmental approvals, which could impact project timelines and operational efficiencies. In cases where TSOs of regulatory bodies are located in different jurisdictions, these complexities are compounded, underscoring the need for clear, adaptable frameworks that respect national sovereignty while facilitating the necessary international regulatory cooperation.

Consequently, the following outlines the associated impediment discussion. It would be difficult or impossible for some States who require export approvals to achieve an open licence to allow the sharing of export-controlled information with other States. Issuing an export approval can involve additional interested parties and might result in complications or delays.

3.1.4. Pre-decisional information

Regulatory bodies conducting collaborative or joint regulatory reviews may want to exchange information that contains a regulatory body's draft or preliminary views of the application. This may include draft safety evaluations or other information that reflects the regulatory body's internal deliberations or initial views that are incomplete, pending further legal or management review, or otherwise subject to further change. Such information may be considered by the regulatory body to be pre-decisional information that is exempt from public disclosure under its governing laws. In general, a regulatory body may consider its own pre-decisional information to be sensitive because its premature release may lead to public confusion or may negatively influence agency deliberations amongst its staff.

Consequently, the following outlines the associated impediment discussion. Generally, there would not be a legal impediment to sharing pre-decisional information. Unless prohibited by its own domestic law or policy, a regulatory body would have discretion to decide whether to share its own pre-decisional information with another regulatory body in the context of a joint review or a collaborative review. However, any such sharing would generally need to be done pursuant to a written agreement or other understanding where the recipient would agree to respect the confidentiality of the information and protect against public release. Additionally, pre-decisional information that includes other categories of controlled information discussed in this section (e.g. pre-decisional discussions of NSR information or pre-decisional analyses that incorporate proprietary information) would need to be controlled in accordance with the requirements governing those other categories, which could be more stringent.

3.2. IMPEDIMENTS WITH AVAILABILITY AND ACCESS TO INFORMATION

3.2.1. Documents supporting regulatory reviews

This subsection identifies potential impediments to sharing documents that may support the regulatory reviews performed by a regulatory body or a group of regulatory bodies during future bilateral or multilateral joint international activities.

As discussed in Section 2, documents that support regulatory reviews can be grouped into two general categories:

- Documents submitted to a regulatory body by the DIO;
- Documents generated by a regulatory body.

3.2.2. Observing detailed design information

With respect to observing information at a DIO's premises, the design information available would typically be at a more granular level of detail than the information contained in the DIO's submission to a regulatory body. Accessing this supporting information, such as detailed PSA analyses, may help the regulatory bodies better understand the information submitted by a DIO and more efficiently conduct collaborative and joint reviews. This will require the regulatory bodies to travel to the DIO's premises.

With respect to observing information virtually, detailed design information may also be observed using the DIO's online electronic portals which are commercially available and commonly utilized in the nuclear industry. The use of electronic portals/reading rooms will likely require prior approval from all of the regulatory bodies and the DIO, in addition to meeting any applicable export controls. This approval will consider conditions on use of password protection (such that passwords will be assigned only to those directly involved in the joint regulatory body review on a need-to-know basis) and preventing or disabling any printing, saving, or downloading functions.

Consequently, the following outlines the associated impediment discussion. It is anticipated that information reviewed at the DIO's premises, or via the DIO's electronic portals/reading rooms, will likely contain controlled information (such as trade secrets, confidential commercial, or financial) as discussed in Section 3.1. Therefore, regulatory bodies will need to put in place mechanisms to control access to and prevent the unauthorized release of any controlled information.

3.2.3. Computer code models and input data

This section identifies potential impediments with regulatory bodies gaining access to a DIO's:

- Computer code models;

- Input data;

- Detailed design information needed as input to a regulatory body's independent computer models.

Consequently, the following outlines the associated impediment discussion. It is anticipated that modelling and simulation tools, computer codes, computational analyses and input data will likely contain controlled information as discussed in Section 3.1. Therefore, approvals and agreements will be needed if regulatory bodies performing future collaborative and joint reviews and leveraging other regulatory reviews of advanced reactor designs want to share modelling and simulation tools, computer codes, computational analyses and input data. Regulatory bodies will need to put in place mechanisms to control access to and prevent the unauthorized release of any controlled information.

3.3. IDENTIFICATION OF KEY STAKEHOLDERS NEEDED TO DEVELOP SOLUTIONS

This section identifies interested parties that need to be involved in developing information sharing agreements and who have an interest in the information sharing. Each interested party's type and level of engagement will depend on their level of involvement in information sharing. Three categories of interested parties have been identified:

- Those involved in sharing and leveraging information;

- Those involved in providing solutions to resolve impediments to information sharing;

- Those with a valid interest in what the States are doing.

3.3.1. Key interested parties

Table 1 identifies key interested parties for the three categories mentioned above.

TABLE 1. KEY INTERESTED PARTIES

Category	Interested parties
Those directly involved in sharing and leveraging information	<ul><li>DIOs, other information owners and applicants or licensees</li><li>Regulatory bodies<ul><li>Federal</li><li>Provincial</li></ul></li><li>TSOs and individual contractors</li><li>Industry associations</li></ul>
Those involved in providing solutions to resolve impediments to information sharing	<ul><li>Government<ul><li>Federal (export control, trade, security, customs and inter-governmental areas)</li><li>Regional</li></ul></li></ul>
Those with a legitimate interest in what the States are doing	<ul><li>Federal government departments (business/industry development, climate change, environment, emergency preparedness, transport subject areas)</li><li>Regional governments/potentially affected local communities</li><li>Other utilities, vendors, suppliers, unions</li><li>Public</li><li>Non-governmental organizations (NGOs)</li></ul>

3.3.2. Interested parties' involvement in regulatory review

The involvement of each interested party will differ as the activities performed by the regulatory bodies involved in joint, collaborative reviews and/or leveraging information progress from initiation to execution and to completion. An example of how involvement could change throughout the review process and what communication and engagement actions are needed at each stage, is represented in Table 2. It is understood that the regulatory bodies performing these cooperative activities are responsible for coordinating and encouraging these actions. For the purposes of Table 2, the following definitions are used:

- Initiation – several regulatory bodies start setting up the sharing mechanisms to conduct a collaborative or joint review or to leverage other regulatory bodies' reviews.

- Execution – the time during which the regulatory bodies are undertaking the collaborative or joint review or leveraging other regulatory bodies' reviews.

- Completion – the collaborative or joint review or the leveraging of other regulatory bodies' reviews has been completed, either in-part or entirely, and conclusions are available.

Careful consideration will be given to engagement of interested parties to ensure that the appropriate level of involvement is sought, at the right time, when attempting to identify solutions and to implement them. A clear engagement plan will need to be drafted to ensure early buy-in from all the relevant parties in advance. These plans will need to incorporate a communications strategy, methods for gathering feedback, and a process by which differences of opinion can be addressed and resolved.

TABLE 2. COMMUNICATION AND ENGAGEMENT AT DIFFERENT STAGES OF REGULATORY REVIEW

Category	Interested parties	Collaborative or joint regulatory review or leveraging information stage		
		Initiation	Execution	Completion
Those directly involved in sharing and leveraging information and developing sharing information agreements for all type of controlled information	DIOs: vendors and other information owners	<ul><li>Confirm the information needed and the timing thereof.</li><li>Confirm who the information will be shared with and why.</li><li>Definc and agree arrangements for sharing controlled information (issue NDAs, or other, to regulatory bodies and interested parties). Note that this includes defining requirements, if appropriate, regarding the recording of information shared and the associated quality assurance.</li><li>Be aware of what agreements (export control, etc.) will be needed and when, taking into account sufficient lead time.</li><li>Grant approval to share information.</li></ul>	<ul><li>Hold regular follow-up meetings with regulatory bodies and other interested parties to confirm level of detail within the information is sufficient.</li><li>Send requests to the vendor for more information resulting in additional information being shared and/or updating of information previously shared.</li><li>Review the arrangements in place and identify opportunities for efficiencies.</li><li>Audit records, as necessary.</li></ul>	Review how information was used and identify opportunities for efficiencies

TABLE 2. COMMUNICATION AND ENGAGEMENT AT DIFFERENT STAGES OF REGULATORY REVIEW (cont.)

Category	Interested parties	Collaborative or joint regulatory review or leveraging information stage		
		Initiation	Execution	Completion
	DIOs: applicants/ licensees	<ul><li>Confirm the need for any site specific or operational information and the timing thereof.</li><li>Grant approval to share information, as necessary.</li></ul>	<ul><li>Hold regular meetings with interested parties to increase knowledge on reactor design and understanding of challenges which may arise in licensing activities.</li><li>Hold regular meetings with the regulatory body and vendor to discuss progress and ensure alignment.</li><li>Review arrangements in place and identify opportunities for efficiencies.</li></ul>	Review how information was used, how useful the output will be in the licensing process and identify opportunities for efficiencies.

TABLE 2. COMMUNICATION AND ENGAGEMENT AT DIFFERENT STAGES OF REGULATORY REVIEW (cont.)

Category	Interested parties	Collaborative or joint regulatory review or leveraging information stage		
		Initiation	Execution	Completion
	Regulatory bodies	<ul><li>Define what information will be needed and when.</li><li>Review DIO NDAs to determine if they are acceptable and/or define processes to protect intellectual property.</li><li>Set up information sharing mechanisms.</li><li>Grant approval to share information, as necessary.</li><li>Put in place the mechanisms for recording and reporting information sharing and the quality assurance.</li></ul>	<ul><li>Manage flows of information between regulatory bodies.</li><li>Hold regular meetings with DIO to update on progress, ensure alignment and identify gaps in information (scope or depth).</li><li>Record, quality assure and report, as necessary.</li></ul>	Gather best practices and identify challenges in sharing information and proposed solutions.
	TSOs and individual contractors	<ul><li>Have infrastructure and agreements in place with regulatory bodies to handle controlled information.</li><li>Set up information sharing mechanisms with regulatory bodies.</li><li>Put in place the mechanisms for recording and reporting information shared and the quality assurance.</li></ul>	<ul><li>Manage flows of information between regulatory bodies and TSO/contractors.</li><li>Record, quality assure and report, as necessary.</li></ul>	<ul><li>Identify best practices and challenges in sharing information and proposed solutions.</li></ul>

TABLE 2. COMMUNICATION AND ENGAGEMENT AT DIFFERENT STAGES OF REGULATORY REVIEW (cont.)

Category	Interested parties	Collaborative or joint regulatory review or leveraging information stage		
		Initiation	Execution	Completion
Those involved in providing solutions to resolve impediments to information sharing that are export controlled and security related	Government	• Understand the types of information to be exchanged and when. • Identify legal and policy difficulties in sharing certain types of information with the relevant States. • Identify and involve interested parties in a timely manner. • Define the export control challenges and timescales for resolution. • Inform the public of proposed plans. • Granting approval to share information, as necessary.	• Solicit feedback and identification of solutions. • Ensure resolution of impediments identified during initiation to allow work plans to advance as planned.	• Identify challenges in resolving information sharing impediments and propose solutions. • Gather best practices, lessons and feedback from engagement with general public.
Those interested in what the States are doing	Government departments not directly involved in regulatory reviews (e.g. business/industry development, climate change)	• Be informed regarding the activities being undertaken, and what information is being shared and why.	• Update public as necessary.	• Provide feedback to participants in information sharing, as necessary. • Update public as necessary.

TABLE 2. COMMUNICATION AND ENGAGEMENT AT DIFFERENT STAGES OF REGULATORY REVIEW (cont.)

Category	Interested parties	Collaborative or joint regulatory review or leveraging information stage		
		Initiation	Execution	Completion
	Regional governments / potentially affected local communities	• Be informed regarding the activities being undertaken and what information is being shared, and why. • Provide feedback and input local concerns.	• Continue to provide feedback and input local concerns	• Gather best practices, lessons and feedback from engagement with regional government and potential future licensees.
	Other utilities, vendors, suppliers, unions	• Understand the types of information to be exchanged and when.	• Gather outputs and feedback in terms of information sharing. Input lessons from previous activities.	• Provide expertise and insight where necessary. • Facilitate lessons learnt/feedback meetings.
	Observers (e.g. public, NGOs)	• Be informed regarding the activities being undertaken and what information is being shared, and why. • Provide feedback as and when requested	• Continue to provide feedback and engage with interested parties as requested.	• Gather best practices, lessons and feedback from engagement process.

4. IDENTIFICATION OF POTENTIAL SOLUTIONS FOR IMPEDIMENTS TO INFORMATION SHARING

4.1. KEY IDENTIFIED IMPEDIMENTS

4.1.1. Summary of impediments

As discussed in Section 3, there are numerous potential impediments to freely sharing controlled information between the regulatory bodies of different States. These can include impediments of a generally applicable nature that exist regardless of the type of controlled information being shared, or unique impediments associated with particular categories of controlled information. Omitting discussions at DIO's premises, which can be subsumed within considerations of proprietary information, the key impediments discussed in Section 3 are:

- With respect to information that has previously been designated as proprietary, States will likely have legal and/or policy obligations to obtain the consent of the owner of that information, the DIO, before disclosing it to others (see Section 3.1.1);

- With respect to information that has been determined by a State to be NSR or related to safeguards, the State may have legal or policy restrictions, or even outright prohibitions depending on the level of detail, preventing the sharing of that information with others (see Section 3.1.2);

- Information concerning technology or software that may be of use in the utilization of special nuclear material or an unsafeguarded nuclear fuel cycle may require additional authorization from appropriate export officials before it can be shared outside the State (see Section 3.1.3);

- Regulatory bodies may be reluctant to exchange pre-decisional information containing draft or preliminary reviews (see Section 3.1.4);

- Regulatory bodies that are able to share controlled information pursuant to written agreements may still encounter various administrative issues, such as those relating to translation of agreements; in addition, disparate domestic frameworks for protecting controlled information may present challenges when negotiating such agreements.

4.1.2. Categories of impediment

In general, the identified impediments can be divided into two categories based on the level of control that the regulatory body possesses to overcome the impediment:

Within regulatory body control: these are impediments to sharing controlled information that the regulatory body can overcome without depending on the approval of another party. Although the regulatory body ultimately retains the discretion of whether or not to share information, in many cases there will still be legal or policy criteria that will need to be satisfied in order for the impediment to be overcome. Examples include:

- Depending on the level of detail, a regulatory body may have discretion to share NSR information, provided that the regulatory body is satisfied that the recipient of the information has a requisite need and can protect the information accordingly. This judgment generally would be at the discretion of the regulatory body.

- A regulatory body generally would have full discretion in determining whether and to what extent to release non-public, pre-decisional information generated by its own employees.

Beyond regulatory body control: these are impediments where the regulatory body's ability to share information with another regulatory body is subject to, or reliant upon, the consent or approval of another party such as the DIO or a governmental entity (see Table 1). Some examples include:

- Commercial information (trade secrets or other confidential-commercial or financial data), or other information to which intellectual property rights apply. Refusal by the owner of that information to permit its distribution to others may impede or outright prohibit the sharing of that information. This is similar to an impediment associated with accessing information at DIO's premises.

- Export-controlled information may also fall into this category, depending on the legal framework of the State. For example, if a regulatory body receives a request to share information that needs export approvals or is subject to other export controls, it may first need to coordinate with and obtain approval or concurrence from other government entities, if the authority for such approvals resides elsewhere within the State's government.

- Some types of NSR information may also be included in this category, depending on its significance or level of detail. For example, in some States the approval from an internal ministry or other appropriate national security authority may be needed before certain kinds of NSR information (e.g. threat assessments or a design basis threat) could be shared.

The interested parties that may be involved in addressing the above impediments are identified in Tables 1 and 2.

4.1.3. Solutions to impediments

4.1.3.1. *Within regulatory body control*

These impediments will generally involve the use of written agreements to facilitate an information exchange. These written agreements will need to include clauses limiting regulatory bodies' commitments to what is achievable to the extent permissible by their own domestic laws and policies. This is because the States may have different domestic laws or requirements governing the same subject matters that will need to be reconciled. Ultimately, in order to exchange controlled information, regulatory bodies will need to gain confidence that any requirements in their own laws concerning the handling or further dissemination of the controlled information will be honoured once it is in the possession of the receiving regulatory body.

For example, a regulatory body may determine that it is legally able to share certain controlled information so long as the recipient agrees to incorporate the same standards for the handling and protection of the information: the recipient State will likewise have to determine whether those standards are consistent with its own domestic laws or policies and the extent to which it can achieve them. The regulatory bodies will need to determine to what extent they can be flexible and find common ground when negotiating these agreements. In some cases, regulatory bodies may have threshold differences of policy or interpretation as to whether certain information qualifies as controlled information. However, assuming it is not contrary to domestic law, the State receiving the information may have the flexibility to agree to treat the information as controlled information and respect the other party's designation, even if that recipient has not made the same designation on its own. Such flexibility may facilitate the exchange of information where differences in domestic law or policy exist.

4.1.3.2. Beyond regulatory body control

These are the most potentially formidable impediments to sharing controlled information, while also potentially the most straightforward to overcome. They are potentially formidable in that failure to obtain the necessary consent or approval from the appropriate party can result in a total prohibition on sharing the controlled information. The State will need to proactively identify which impediments in their respective domestic laws and policies necessitate the approval or consent of another party, and then actively engage with those parties in a timely manner to ensure their ability to share controlled information (as illustrated in Table 2). However, in some cases obtaining this consent or approval from the appropriate party will not be difficult, because sharing the controlled information will be in that party's interests or even initiated by the appropriate party itself.

For example, a DIO that expresses an interest in initiating a joint review, such as the multinational pre-licensing review process being developed by NHSI Working Group 2 [3], will first provide advanced notification of their plans to regulatory bodies participating in the joint review process. Then, this DIO will need to provide consent to allow each participating regulatory body to access and freely share amongst each other any commercial information the DIO submits for review. In this process a DIO would only select regulatory bodies that it found acceptable in terms of obtaining its proprietary information. Additionally, when two or more regulatory bodies are engaged in collaborative reviews of the same design or leveraging information from a previously completed review by another regulatory body, as envisioned by the processes developed by NHSI Working Group 3 [2], it will be in the best interest of the DIO to provide consent for the regulatory bodies to share information so that they might conduct their review of the DIO's information.

Regulatory bodies could possibly avoid or minimize these impediments by having the DIO provide, to the extent possible, controlled information directly to another regulatory body. Depending on the circumstances this could be preferable from the DIO's perspective, as it will have more control over what is shared and will be in direct contact with those who possess its information, rather than being an outside party to a regulatory body-to-regulatory body exchange of its information. This might not always be feasible, however, for example where regulatory bodies incorporate controlled information into their own documents and analyses that they share with one another.

4.1.4. Additional administrative impediments

The categories of impediments described above relate to controlled information for which law, regulations, or policy restrict its free exchange. In addition to these impediments, regulatory bodies exchanging information could also face administrative impediments not strictly imposed by any specific domestic law, regulations, or policy. Examples include determining the resources or the amount of assistance a State might provide to translate documents, or determining the administrative format that a written agreement would embody (e.g. a binding or non-binding agreements, which could include memoranda of agreement or terms of cooperation, contracts or non-disclosure agreements, etc.). These types of administrative impediment can be discretionary preferences that regulatory bodies have when exchanging information, in which case they can generally be overcome through mutual agreement, although the process and length of negotiating such an agreement can delay the exchange of controlled information.

4.2. REGULATORY COOPERATION INVOLVING AN INFORMATION SHARING AGREEMENT

It is expected that any exchange of controlled information between regulatory bodies will involve the use of written information sharing agreements. These agreements would include certain clauses intended to ensure the protection and confidentiality of controlled information. Clauses that may be included in such written agreements are described in Section 4.3. However, there are many different types of cooperation being contemplated as part of NHSI (see Section 2 and Annex III), and the scope of these written agreements and their structure could vary depending on a number of factors, such as:

- The number of parties to the written agreement (e.g. bilateral versus multilateral);

- Whether the parties to the written agreement are limited solely to regulatory bodies, or whether the agreements include other parties such as DIOs and TSOs;

- Whether the parties to the written agreement intend to utilize another organization (e.g. IAEA) to provide secretariat or repository functions;

- Whether the collaboration is initiated by the DIO or by regulatory bodies;

- Whether the cooperation involves regulatory bodies individually reviewing the design against their own domestic legal frameworks (collaborative review) or are participating in a joint review against another agreed-upon framework.

Sections 4.2.1–4.2.3 summarize the three different types of cooperation that are under consideration as part of NHSI, and propose an information sharing framework for each that may facilitate a more efficient exchange of controlled information (see also Annex III). For joint and collaborative review projects, there would need to be broader agreements in place governing the scope of the project, including specifying applicable information sharing clauses.

This subsection does not address any contractual agreement or information sharing arrangement that is solely between a regulatory body and a DIO. It is presumed that the submission of information to a regulatory body by a DIO will be conducted pursuant to the domestic laws, regulations and processes of the regulatory body's State, and that these domestic laws, regulations and processes will govern the rights and obligations of the parties. Rather, this section focuses solely on scenarios in which a regulatory body, in cooperation with a DIO, shares controlled information with one or more regulatory bodies from other States; this information could have been received directly from a DIO or it could have been generated by the regulatory body itself.

4.2.1. Joint reviews

There are several approaches for regulatory bodies to undertake a joint review of an advanced reactor design. This subsection is focused on the DIO-initiated joint review scenario described in Ref. [3].

In this scenario, a review team comprised of nuclear safety staff from regulatory bodies from multiple States would collaborate and provide a joint, non-binding opinion, on the acceptability of a proposed advanced reactor design against an established reference framework. This framework may be based on an amalgamation of various national, international or regional regulatory requirements. To initiate this process, a DIO would first engage in informal bilateral discussions with the appropriate regulatory bodies in States in which the DIO is planning to deploy the design. Each regulatory body would then decide whether it was willing to participate in a multinational pre-licensing joint review of the design. If enough interest is expressed, the review process would be initiated and established within bilateral or multilateral agreements (see Annex III) in alignment with the scenarios considered in Ref. [3]: (a) through an international organization such as the IAEA which would provide a secretariat function, receive the DIO's design documents and manage access; or (b) through the DIO itself, by directly providing each individual regulatory body with access to the necessary design documents.

This type of joint review may present relatively fewer impediments to establishing an appropriate information sharing framework than other scenarios. This is because it is a DIO-initiated process involving only regulatory bodies that have already been selected by the DIO. Thus, the impediments beyond regulatory body control described in Section 4.1 will already be addressed at the initiation of the joint review. It is not expected that regulatory bodies would need to obtain additional approvals from the DIO in order to exchange or discuss DIO's proprietary information that has already been shared equally with all the participants in the review. Additionally, in this scenario the DIO would be providing its design information to pre-selected regulatory bodies (either directly or through an intermediary organization), and thus in order for this scenario to be initiated the DIO would need to select only regulatory bodies located in States where government-to-government assurances are already in place, to enable the necessary export licences to be obtained. Thus, similar to proprietary information, the exchange of export-controlled information in this scenario would not be expected to be an impediment. And because each regulatory body would be provided with access to the DIO's information, it is unlikely that regulatory bodies would need to share or exchange copies of DIO documents in this scenario (though regulatory bodies may generate their own documents that incorporate DIO proprietary information, and these would need to be exchanged subject to information protection controls) and export approvals.

In the joint review scenario, where a DIO is initiating a specific project to produce a single joint review document, a single written agreement would likely suffice to enable the sharing of controlled information amongst all participants. Inherently, the presence of established nuclear cooperation among States makes negotiating and implementing this agreement easier to achieve, especially when few participants are involved. The primary purpose of this written agreement would be to document the DIO's consent to the sharing of its proprietary information, as well as documenting the parties' obligations to protect the confidentiality of such information when in their possession. The written agreement would include all parties involved in the review, including (a) the DIO; (b) each participating regulatory body; and (c), if applicable, the organization providing the secretariat or information repository function. This single written agreement could incorporate the information protection clauses described in Section 4.3, as appropriate, to address information protection when regulatory bodies exchange information or documents. The DIO information, while in the possession of a regulatory body, would remain subject to the regulatory body's domestic laws, regulations and policies.

4.2.2. Collaborative reviews

There are several approaches for regulatory bodies to undertake a collaborative review of a design of an SMR or other advanced reactor. This subsection is focused on the DIO-initiated collaborative review scenario described in Ref. [2].

In this scenario, a DIO would provide a group of regulatory bodies with individual access to design information to enable each regulatory body to review the design, in parallel, against their own domestic legal frameworks. The DIO would provide each regulatory body with access to the information, and, with DIO's consent, these regulatory bodies would be permitted and encouraged to collaborate and share information with one another in the course of their parallel, independent reviews.

Similar to the joint review scenario described in Section 4.2.1, this scenario would not be expected to raise significant impediments with respect to information beyond regulatory body control described in Section 4.1. Because this scenario is also a DIO-initiated scenario, only regulatory bodies that have been selected by the DIO as appropriate recipients would receive access to proprietary information. Thus, it would not generally be expected that regulatory bodies would need to seek or obtain additional approvals from the DIO in order to share or discuss DIO proprietary information, since the same information has already been made available to the other regulatory bodies conducting their independent reviews. However, challenges could emerge if one regulatory body requested additional controlled information from the DIO that was not provided to all regulatory bodies conducting their own independent review of the design. If all regulatory bodies do not have equal access to the same DIO information, then regulatory bodies would need to take additional steps to confirm with the DIO that information not commonly shared with all regulatory bodies could be shared with other regulatory bodies. Thus, in the collaborative review scenario, DIOs would need to be encouraged to share with all regulatory bodies any additional information provided to one, to better ensure a more seamless ability for regulatory bodies to communicate and collaborate. If a DIO elects not to do so, it will make clear to the regulatory body being provided with the additional information that it cannot be shared with others, so the regulatory body knows that information will be excluded from any cooperative efforts.

Although written agreements between regulatory bodies might not be necessary in this scenario, to facilitate collaboration involving DIO proprietary information, written agreements would still be necessary in order for regulatory bodies to exchange other types of controlled information, such as NSR information or pre-decisional information. This is because in this scenario, regulatory bodies would be reviewing the DIO's information against their own domestic legal frameworks, and thus analyses and documents generated by a regulatory body could potentially include these types of controlled information (e.g. assessments involving NSR considerations of threat assessments, or preliminary or draft regulatory reviews that are not made publicly available). These written agreements could be made on a bilateral basis or multilateral basis including all regulatory bodies participating in the parallel collaborative reviews. The written agreements could incorporate clauses described in Section 4.3, as appropriate.

4.2.3. Leveraging reviews

Another type of NHSI collaboration that might be envisioned is for regulatory bodies to share information obtained or created in the course of their own domestic regulatory review processes. This may include requests from other regulatory bodies that are reviewing the same DIO design at the same time (i.e. leveraging an ongoing review), or a regulatory body leveraging information from another regulatory body that has previously performed its own review of a specific design (i.e. leveraging a completed review). These scenarios differ from those discussed above because they are requests for information or cooperation that are not initiated by the DIO. Scenarios involving regulatory bodies leveraging the reviews of other regulatory bodies is discussed in Ref. [2].

This scenario is likely to result in more impediments to exchanging information. Unlike DIO-initiated scenarios, regulatory bodies receiving requests for information might not have the requisite assurance at the time of the request that information beyond regulatory body control (as described in Section 4.1, and including proprietary information or export-controlled information) is authorized for exchange. Unless the regulatory body has such assurance, it will need to contact the appropriate information owner to obtain approval to share with the requesting regulatory body. Depending on the State, the regulatory body could also need to obtain assurance from the DIO or approval from the appropriate government authority that sharing the information with the requesting regulatory body is approved from an export control perspective.

For the same reason as described in Section 4.2.2, regulatory body-initiated requests for information or cooperation will involve written agreements in order to exchange controlled information. Additionally, unlike DIO-initiated scenarios described above, a regulatory body-initiated request for information is more likely to occur on an ad-hoc or unanticipated basis. Early receipt of such requests will greatly help to achieve a positive response. As explained in Section 4.3.2, one solution to potentially mitigate challenges to entering into these agreements is for States to first enter into a non-binding, multinational MoC that would affirm the mutual interest of participants in cooperating in the interests of global nuclear safety. Specific details governing the exchange of controlled information could then be left to subsequent implementing agreements between individual or smaller groups of regulatory bodies.

4.3. PROVISIONS FOR INFORMATION SHARING BETWEEN REGULATORY BODIES

This subsection discusses provisions that regulatory bodies may consider for inclusion in written agreements for the sharing of controlled information. It is recognized that regulatory bodies may already have information sharing agreements in place with other regulatory bodies. If these pre-existing written agreements are sufficiently broad or flexible enough to facilitate the exchanges of information described in this publication, then regulatory bodies may use those agreements in lieu of creating new information sharing agreements.

Regulatory bodies that have obligations to obtain approvals from other parties in order to share controlled information will need to do this prior to entering into a written agreement. As discussed in Section 4.2, it is expected that the relevant approvals will already be secured at the outset of a DIO-initiated scenario, although this might not necessarily be the case in a regulatory body-initiated scenario. If all requisite approvals are obtained, then regulatory bodies could use these written agreements to ensure that any domestic legal or policy requirements governing controlled information will be implemented by the recipient of that information. If permissible under their domestic legal frameworks, regulatory bodies may need to accept that information shared with other regulatory bodies might not be handled in exactly the same way and be prepared to accept risk-informed solutions or compromises that provide an acceptable level of protection based on the circumstances.

Ultimately, regulatory bodies entering into these implementing agreements will be responsible for negotiating their contents and deciding on their format and terms, but it is expected that such implementing agreements would, at a minimum, need to address the areas discussed in Sections 4.3.1 and 4.3.2.

4.3.1. Written agreements

4.3.1.1. Identification and marking

It is imperative that information sharing agreements for the exchange of controlled information obligate the sender of that information to expressly notify the recipient of its controlled status at the time the information is exchanged, and prominently mark it as such in a way that is recognizable by the recipient. While regulatory bodies may have their own specific requirements under domestic law or policy for how controlled information is to be marked while in their possession (e.g. specific text or acronyms to be applied in headers and footers, stamps, watermarks, etc.) these State-specific mechanisms for identifying controlled information might not be readily understood when exchanged with other regulatory bodies subject to different requirements. Thus, parties to an information sharing agreement need to agree on an appropriate language or marking scheme that can accompany all controlled information that is exchanged, ensuring that both parties have a mutual understanding of the status of the information. Assuming it is not contrary to a State's domestic law or policy, such markings could be incorporated into information in addition to (and not in lieu of) any State-specific marking requirements.

Information sharing agreements will also expressly address the situation of a recipient of controlled information copying, reproducing or even, perhaps, referencing that information in a derivative document and specify that any such copy or reproduction be marked and handled consistently with the original source of information (i.e. specifying that controlled information does not lose its controlled status when incorporated into a new document).

4.3.1.2. Restrictions on further dissemination

The regulatory bodies entering into implementing agreements for the exchange of controlled information will specify the extent to which that information can be shared or further disseminated, both within and outside the recipient regulatory body. Depending on the nature and sensitivity of the information being exchanged, regulatory bodies may decide to limit access to:

- Specific named individuals employed by the regulatory body;

- All employees within the organization who have a requisite need to know the information to perform official duties;

- Any employee of the regulatory body's government with a requisite need to know;

- Individuals outside of the regulatory body but who support the regulatory body and have a requisite need to know the information to provide such support (e.g. contractors, advisory bodies, TSOs, etc.).

It is not expected that individual NDAs would be necessary for exchanges limited solely to employees of a regulatory body, since such employees will generally already be prohibited by their domestic laws or policies from making unauthorized disclosures of non-public information. However, NDAs may be necessary or advisable if controlled information is permitted to be shared with individuals not employed by the regulatory body, such as contractors or TSOs. In lieu of individual NDAs, regulatory bodies may seek assurances that such support individuals are governed by contracts or otherwise have legally enforceable obligations to protect confidential information that are at least as extensive as that of the regulatory body. Regulatory bodies may also need to expressly reserve the right to approve any such dissemination outside the regulatory body, to the extent permitted by their domestic laws. This is especially important where dissemination of the information is subject to the approvals of another party (such as proprietary information), because dissemination outside the regulatory counterpart may be beyond the scope of what the other party has previously authorized.

4.3.1.3. Restriction with respect to purpose

Regulatory bodies entering into information sharing agreements may also specify that information exchanged pursuant to the written agreement is to be used only for the purpose of the cooperative activity and not for any other purpose without the approval of the regulatory body that provided the information.

4.3.1.4. Information security

States have their own domestic laws, regulations and policies governing the security of controlled information that they could seek to incorporate into written agreements. This includes:

- Physical security requirements to guard against unauthorized disclosure when information is in hard copy format (e.g. requirements for proper storage, physical access controls, or limitations on use only in appropriately controlled environments).

- Digital security requirements, such as restrictions on downloading or printing controlled information, or minimum acceptable cybersecurity standards or assurances. Regulatory bodies that have disparate cybersecurity postures may need to explore options such as electronic reading rooms or information portals with read-only access (i.e. no download or copying capability) in order to facilitate exchanges of sensitive controlled information in digital format.

- Requirements for the return or acceptable destruction of controlled information when no longer needed.

With respect to information security requirements, States may incorporate specific standards into their written agreements that are either prescriptive or performance-based, depending on their own domestic laws and policies. Where regulatory bodies have flexibility, performance-based standards may be preferable and more conducive to facilitating an exchange. Both parties to an exchange may more readily be able to determine that performance-based standards are in conformance with their own domestic laws and policies, as opposed to specified prescriptive standards that might create conflicts. As an illustrative example, a written agreement could specify that controlled information will be destroyed by any method that renders the information unreadable and indecipherable, as opposed to specifying a specific particular destruction standard (e.g. shredded to a certain size) that may be different from the standard used by the other party.

4.3.1.5. Timely notification of spill or breach

Written agreements for the exchange of controlled information will impose obligations to promptly notify the other party of any suspected or confirmed unauthorized disclosure or 'spill' of controlled information received from the other party. This notification needs to be provided regardless of fault. The regulatory bodies can decide in their written agreements whether the parties will have obligations beyond prompt notification in the event of an unauthorized disclosure. This for example may include specific arrangements for record keeping, quality assurance and reporting.

4.3.1.6. Public transparency laws

Many States have public transparency or freedom of information laws that make records created by, or in the possession of, governmental entities publicly available by default, unless a recognized exception to the law permits non-disclosure. The regulatory bodies will likely need to insert clauses into their written agreements to afford the maximum amount of protection under these laws to controlled information that is exchanged.

First, the regulatory bodies may need to insert clauses into their written agreements that expressly state that information exchanged pursuant to the agreement is to be held in confidence by the recipient and that, by releasing the information to the other party, the regulatory body is not waiving any legal defences or the applicability of any exceptions in its public transparency law in order to be able to withhold that same information in the event of a public request.

Second, the regulatory bodies will likely seek to include assurances in written agreements that recipients of controlled information will withhold such information from the public to the maximum extent permitted by domestic law (i.e. that the regulatory body will not make a discretionary release of such information and will protect it from public disclosure unless compelled by its own domestic law). In the event of a compelled disclosure under domestic law, the written agreement will also specify that the party being compelled to disclose the information will provide advance notification to the party who provided the information.

4.3.1.7. Language and translation

The regulatory bodies may wish to specify in their written agreements an agreed-upon language for the conduct of any cooperation or collaboration. Additionally, because regulatory bodies may be exchanging documents written in different languages, the written agreement may specify the respective responsibilities of the parties to provide or assist with any translation.

4.3.1.8. Other administrative clauses

Other examples of clauses expected to be part of a written agreement for the exchange of controlled information include:

- Warranties and liability – written agreements may specify that information exchanged is provided as-is, or that the party transmitting the information does not provide any warranty as to its suitability for any particular use or purpose, and that reliance upon or conclusions drawn from such information is at the party's own risk and does not give rise to any liability of the other party.

- Costs and resources – written agreements may specify that all costs associated with the exchange of controlled information are to be borne by the party assuming those costs (unless otherwise agreed), and that parties' obligations under the written agreement are subject to the availability of resources and dedicated funds.

- Duration and amendment – written agreements may specify whether they are intended to be open-ended agreements or whether they are to be terminated after a specified term (e.g. number of years) or at the occurrence of a particular event (e.g. final publication of a joint work product). Written agreements may also specify the administrative provisions by which they can be renewed or amended (typically upon mutual written agreement of the parties).

- Termination and continuing obligations – written agreements may also address the administrative mechanism by which one party to the agreement can terminate participation (typically after a prescribed period of advance notice). Termination clauses may also address any remaining obligations upon termination (for example, requiring the return or destruction of controlled information in the possession of the terminating party), including that, in the event of the written agreement being terminated, the parties' respective obligations to protect controlled information received from the other party continue to apply.

- Dispute resolution –the parties may determine an appropriate course of resolution for disputes under the written agreement,. This could include informal dispute resolution (e.g. a commitment to amicably resolve disputes between the parties, with elevation to senior government officials if necessary) or a formal dispute resolution mechanism, such as arbitration by mutual agreement in an appropriate forum.

For the avoidance of doubt, written agreements may also include clauses commonly referred to as 'reservation' clauses or 'savings' clauses, specifying that nothing contained within the agreement is to be interpreted as to require a party to take any action that would be inconsistent with that party's domestic laws, regulations, or policies.

4.3.2. Memorandum of cooperation

It is unlikely that all States interested in future collaboration as envisioned by NHSI will be able to enter into one large multilateral agreement that is legally binding and would facilitate the exchange of controlled information between all signatories. Rather than one large multilateral agreement that establishes the governance for the exchange of controlled information, a non-binding MoC is the most feasible way for regulatory bodies representing a large number of States to enter into a common written understanding concerning the potential exchange of such information. This MoC would be an aspirational document that expressly affirms the mutual interest of regulatory bodies willing to cooperate with one another in the interest of global nuclear safety, consistent with their national roles and responsibilities. In this MoC, regulatory bodies could broadly commit to exchanging information to the fullest extent permitted by their domestic laws and national policies but would specify that the exchange of controlled information is to be made pursuant to subsequent implementing agreements between specific regulatory bodies who are cooperating or collaborating on a specific project. Consequently, the MoC would establish an understanding and an expectation for interested regulatory bodies to make bona fide efforts to cooperate with one another where practicable, but the specific details governing the exchange of controlled information would be left to the more detailed implementing agreements between individual or smaller groups of regulatory bodies described in Section 4.3.1. It is recognized that the ease or ability of any individual regulatory body to enter into this MoC, notwithstanding that it would be non-binding, will greatly depend on each State's individual domestic legal framework, including approval from an internal ministry or other elements of national government. An example of a proposed MoC that may be used for this purpose is provided in Annex I of this publication. Its provisions may be adjusted based on user preferences or their respective domestic legal requirements.

4.4. POSSIBLE APPROACH TO SUPPORT REGULATORY BODIES IN ESTABLISHING THEIR REGULATION OF ADVANCED REACTOR TECHNOLOGY

Section 2.5 identifies additional information needs for regulatory bodies from States that do not yet have a well-established regulatory framework, or from embarking countries, particularly if they are interested in advanced reactor technology for which there is less experience in the application of current safety approaches. There is however a wide variety of information already in the public domain on regulatory reviews of advanced reactors, especially SMRs, by States with well-established nuclear programmes, which could support a regulatory body looking to improve its capability to regulate advanced reactors. Even if the information is in the public domain, it is often not easy to find or to use by embarking countries that are not familiar with the regulatory frameworks of more experienced States and the information may be in different languages.

To support embarking countries and encourage the international collaboration of those interested in potentially deploying advanced reactor technology, an entity could provide an online portal providing the link to the information needed by embarking countries. For example, the following information that can be currently found in various websites of regulatory bodies or vendors could be made more accessible to regulatory bodies in embarking country:

- Regulatory framework for advanced reactor review, including:

  o Changes in regulatory frameworks to ensure applicability to advanced reactors;

  o The application of prescriptive and goal setting risk-oriented approaches;

  o Guidelines and review plans for regulatory review advanced reactor technology.

- Competency requirements and human resource development programme for the regulatory review of advanced reactor technologies.

- Published advanced reactor regulatory review reports and submissions from vendors and/or applicants and licensees, including information on:

  o The general safety aspects of the advanced reactor technology;

  o Identification of all credible initiating events and accident scenarios;

  o The acceptance criteria that were applied;

  o Review of testing facilities and the results;

  o Review of computer codes results or by separate calculation.

- Procedures and guidance on managing and sharing of controlled information, which would be a prerequisite for multinational cooperation. This may include guidance on marking and labelling of controlled information. Also, this may include generic information on handling, storage and sharing of physical and digital forms of controlled information.

- Existing memoranda of understanding or information exchange arrangements between States that can facilitate the sharing of advanced reactor review information.

Understanding how States currently collaborate on the review of advanced reactor designs could help embarking countries to establish information sharing agreements with other relevant parties. With knowledge of existing international advanced reactor regulatory activities (e.g. IAEA's Regulatory Cooperation Forum[2]), embarking countries could learn from completed reviews, where appropriate, it might engage with a review that is still in progress, or else might leverage already completed reviews. Reviewing existing advanced reactor regulatory frameworks would also allow embarking countries to benchmark their current regulatory frameworks against those from experienced States and potentially identify any areas for improvement that would have a positive influence on nuclear safety and security. By facilitating access to the above information, the outcome could be a greater level of international collaboration regarding advanced reactor technology reviews.

A useful source of information that already exists is the IAEA Advanced Reactors Information System which has published 4–5 page descriptions of actual and proposed advanced reactors using a standard format. These descriptions have been supplied by the respective DIOs. The most recent edition (2024)[3] presents data from more than sixty different designs along with global summaries that show the full range of properties and functions. This can be used by embarking countries to understand the range of advanced reactor possibilities; however, an additional tool is needed to better understand the safety and regulatory aspects of these technologies, encompassing the information outlined above.

---

[2] https://www.iaea.org/about/regulatory-activities-section/regulatory-cooperation-forum

[3] https://aris.iaea.org/default.html

4.5. EXAMPLES OF PAST COOPERATIONS AMONG REGULATORY BODIES

The section provides seven examples of cooperations between regulatory bodies regarding reactor design and safety. Five address collaborative reviews as part of pre-licensing reviews of SMR or large scale reactors while two are concerned with leveraging previous reviews for the licensing of large scale reactors (of Russian Federation and ROK design) that may be, or have been, supplied to other States.

4.5.1. SMR Regulators' Forum

The SMR Regulators' Forum[4], hereafter referred to as the Forum, provides enabling discussions among participating IAEA Member States and other interested parties to share SMR regulatory knowledge and experience. The purpose of the Forum is to identify, enhance understanding of, and address key regulatory challenges in emerging SMR regulatory discussions in order to help enhance safety, security and efficiency in SMR regulation, including licensing, and enable regulatory bodies to inform changes, if necessary, to their requirements and regulatory practices.

The Forum is a regulatory body-to-regulatory body entity driven by its members. The outcomes of the Forum do not replace national regulatory bodies' responsibility for making regulatory decisions.

Membership to the Forum is open to all IAEA Member States who are able to contribute in a meaningful manner to the work of the Forum's working groups. Membership requires a commitment to devote adequate resources, with appropriate experience and knowledge base, to ensure that the Forum's purpose and objectives can be fully achieved.

International and regional organizations that express an interest may be granted the status of observer by decision of the Forum's Steering Committee. IAEA Member States can participate as observers to a single combined meeting of the Steering Committee and working groups prior to submitting a request to join the Forum as a member.

4.5.1.1. Governing documents and agreements

Terms of Reference (ToR). Being a member of the Forum requires a dedication to allocate sufficient resources, along with relevant experience and knowledge, to guarantee the realization of the Forum's purpose and objectives.

Communications (internal/external). The Forum utilizes a secure and password-protected online platform within the Global Nuclear Safety and Security Network, managed by the IAEA. This platform serves as a space for information exchange among Forum experts and facilitates communication between the IAEA Secretariat and the Forum. The distribution of pertinent information to external interested parties, including the public, is managed through the Forum webpage.

4.5.1.2. Collaboration publications and outcomes

At the end of each three year phase of work, the Forum issues reports containing common positions on key issues pertaining to SMR regulation. These are made publicly available on the Forum's webpage.

The Forum also provides the outcomes of its discussions through regional educational workshops held twice a year in different regions for regulatory bodies in embarking countries,

---

[4] https://www.iaea.org/topics/small-modular-reactors/smr-regulators-forum

and through webinars. The Forum also identifies lessons learned for enhancing the means and the structure for future cooperation initiatives.

4.5.1.3. Information shared

The Forum mainly discusses publicly available design information. However, for the Forum to be successful in fulfilling its goal of leveraging the work of peer regulatory bodies in the review of SMRs, a framework was developed to facilitate the sharing of technical information among Forum experts. At times, this may include sharing of controlled information such as pre-decisional information following the relevant agreements and policies of the Member States that are part of the Forum.

As a rule, the information exchanged at the Forum meetings and via the Global Nuclear Safety and Security Network restricted website is for the sole use of the participating national regulatory authorities. A large portion of the information shared might not be controlled; however, all participating experts are committed to protect and properly handle the information that a DIO claims to be proprietary.

In August 2022, the Forum officially joined NHSI to lead NHSI Working Group 3. Information sharing among the members of NHSI Working Group 3 is managed through the IAEA Global Safety Assessment Network.

The Forum also regularly exchanges information with its permanent observers: the Nuclear Energy Agency of the Organisation for Economic Co-operation and Development (NEA/OECD), World Nuclear Association Cooperation in Reactor Design Evaluation and Licensing and the European Commission SMR Partnership. The observers provide regular updates on their activities at the Forum biannual meetings and are invited to provide comments on its reports.

4.5.1.4. Repositories

The Forum's publications and generic information for external entities, including the public, are provided on the Forum's webpage.

4.5.1.5. Lessons learned

Key lessons learned regarding information sharing are:

- It is important to establish clear communication channels among Forum experts, between the Forum and the Secretariat, and between the Forum and external interested parties. These channels include dedicated information platforms, such as a webpage used for public communications, and the restricted website, where experts can securely share technical documents, drafts and other relevant information, along with a regular newsletter, web stories and presentations at external events.

- Feedback mechanisms allow the Forum's Steering Committee and its Scientific Secretariat to evaluate the effectiveness of information sharing initiatives and make continuous improvements.

- Having regular meetings provides opportunities for real-time exchange of information and address emerging issues collectively (during topical sessions, special technical sessions and working group discussions).

- Development of publications without controlled information that can be shared widely in open events such as webinars and regional educational workshops contribute to regulatory capacity building via exchange of experiences, lessons learned and best practices.

4.5.2. Multinational Design Evaluation Programme

The Multinational Design Evaluation Programme (MDEP) is a multinational initiative undertaken by national regulatory authorities to develop innovative approaches to leverage the resources and knowledge of the regulatory bodies involved in the review of new nuclear power reactor designs. The NEA/OECD facilitates MDEP activities by providing technical secretariat services for the programme.

The objective of MDEP is to enable increased cooperation within existing regulatory frameworks, and to establish mutually agreed positions that will improve the safety of new reactor designs. MDEP also aims to increase the effectiveness and efficiency of the regulatory reviews, which are part of each State's licensing process.

A key concept of MDEP is that national regulatory bodies retain sovereign authority for all licensing and regulatory decisions.

4.5.2.1. *Governing documents and agreements*

In the beginning of the programme, in order to participate in MDEP and facilitate information sharing, each regulatory body had to sign a bilateral MoU with each of the regulatory bodies from the other MDEP member countries. There was not a standard template for the MoU, it was driven by each country. Currently, bilateral MoUs are not necessary to enter the programme, instead, the regulatory body needs to sign Appendix 3 of the ToR. The MDEP ToR is comprehensive, covering the complete MDEP programme, and provides details about the scope, objectives, programme plan and other items, as discussed below.

The scope includes collaborative efforts, concentrated on reactor design evaluations and regulatory supervision throughout manufacturing, construction, commissioning and early phase operation. These efforts are enhanced through specific technical cooperation within established regulatory frameworks, achieved by establishing shared regulatory stances.

The objectives of MDEP are multifaceted. Firstly, the aim is to enhance cooperation among participants and establish reference regulatory approaches to enhance the safety of emerging reactor designs. Secondly, the focus is on strengthening the effectiveness and efficiency of regulatory reviews. The third objective centres around fostering collaboration on regulatory practices with the ultimate goal of achieving harmonization of regulatory requirements.

In the realm of non-disclosure and proprietary information exchange, a comprehensive protocol is outlined in Appendix 3 of the MDEP ToR. This protocol facilitates the sharing of proprietary design information, thereby unlocking the full benefits of information exchange within MDEP. This exchange extends to its design specific working groups (DSWGs) and issue specific working groups (ISWGs).

ToRs have also been established for DSWGs, representing working groups dedicated to each new reactor design. These groups actively share information and cooperate on specific aspects such as reactor design reviews, construction, commissioning oversight and early phase operation. Furthermore, ISWGs, functioning as working groups for the technical and regulatory process areas identified by the DSWGs, have their own ToRs encompassing high level guidelines for membership, function and objectives, organization and implementation, as well as information management. The ToRs for the DSWGs and ISWGs include:

- High level guidelines, including criteria for membership, for the working groups;

- Function and objective;

- Organization and implementation;

- Information management.

The programme plan is developed and periodically updated by the DSWGs and governed and implemented by the MDEP Management Board. The programme plan describes the long term and short term goals of the working group, the actions that the working group will take to achieve those goals, and a list of products and schedules for those projects.

Non-disclosure agreements are an important aspect of the MDEP. Bilateral agreements exist between most States participating in MDEP and are the most efficient means to set up the framework for a free exchange of proprietary design information between participants. However, since some States do not have such agreements in place, Appendix 3 of the MDEP ToR contains a non-disclosure and information exchange protocol that is made between the MDEP States to support sharing of proprietary design information. This appendix defines proprietary information as trade secret information of any kind, including trade secret information of a business, planning, marketing, or technical nature, provided that such information may have economic value and has been held in confidence. This appendix includes the protocol for sharing proprietary information only on a need-to-know basis with those participants (or their technical support organizations) implementing the particular MDEP activities.

4.5.2.2. Collaboration publications and outcomes

Cooperation among national regulatory bodies under MDEP has led to increased harmonization of regulatory positions and practices through the establishment of common positions, achieved through the activities of different design and issue-specific working groups. Publicly available documents related to MDEP, including final DSWG products, such as Common Positions and Technical Reports, are available on the MDEP webpage[5].

Design-specific common positions are unanimously agreed conclusions reached by the working group participants in the course of the design review. Discussions among the participants and the sharing of information in these areas helped to strengthen the individual conclusions reached. They cover a range of technical topics.

Generic common positions are not limited to one specific design. They are intended to provide guidance to regulatory bodies in reviewing new or unique areas and have been shared with the IAEA and other standards organizations for consideration in standards development programmes.

In addition, the MDEP produces technical reports that enable member countries to better appreciate similarities and understand the differences in national requirements and practices. The topics for technical reports are selected based on the issues arising from regulatory activities in member countries, safety implications, or the general need to have a better understanding of the topic. This facilitates the sharing of experiences and is considered to be the basis of work on harmonization.

---

[5] https://www.oecd-nea.org/mdep/

4.5.2.3. Information shared

The primary focus of MDEP DSWGs lies in the sharing of readily available public design information. Nevertheless, for MDEP to effectively harness the expertise of regulatory bodies in evaluating new NPP designs, it has developed a structured framework facilitating the exchange of technical insights among participants. This framework may entail the sharing of controlled information, particularly proprietary data, as discussed in Section 4.5.2.1. In addition, to date there has been no sharing of export-controlled information and therefore no need for a licence.

As a general rule, the information exchanged at MDEP meetings and via the MDEP library is for the sole use of the participating regulatory bodies. The working group participants also follow the communications protocol and the non-disclosure and proprietary information exchange protocol to share new information related to new reactors with other participants before its release to the public. The main goal of such a communications protocol is to establish a standardized process. MDEP members will give advance notice to relevant parties about any individual or joint reactor design regulatory position statement before making it public. This would allow interested parties to consider the outcome of the review and prepare their responses.

A large portion of the information shared is not controlled; however, all participating members have to protect and properly handle the information that an originator claims to be proprietary, in line with the conditions in the ToR/NDA.

4.5.2.4. Repositories

MDEP information is communicated among the members through the MDEP library, which serves as a central repository for all documents associated with the programme. The MDEP Management Board, through the Technical Secretariat (i.e. the NEA), manages the maintenance of the MDEP library and makes enhancements to improve its effectiveness.

The NEA provides the technical support for development and maintenance of the MDEP library on a secured password-protected website. The website provides two levels of access which are: (a) general access open to every member; and (b) one restricted access area for each MDEP working group that is only open to regulatory bodies that have been or are currently participating in that specific group. For the previous DSWGs, all relevant documentation (minutes of meetings, common positions, technical reports, etc.) have been archived in the MDEP library and are available for future reference.

The NEA Secretariat can use its discretion for requests to access non-proprietary information held in the library. For requests relating to proprietary information received from non-MDEP countries, the NEA Secretariat will either request permission from the originating country to allow access or direct the requestor to the originating country. This ensures that the benefits of MDEP products can be realized by a much wider group of regulatory bodies.

4.5.2.5. Lessons learned

MDEP cooperation among regulatory bodies strengthens the effectiveness and efficiency of the regulatory design reviews, which are part of each country's licensing process. The focus is on cooperation on regulatory practices that will culminate in harmonization of regulatory requirements. In this process, MDEP identified several lessons learned that are specific to information sharing. The MDEP Phase 1 Summary Report for 2006–2021 [15] included the following summary of the lessons learned with information sharing:

"MDEP has made improvements in communicating information regarding member regulatory practices. The development of an MDEP Library has been a major step forward, serving as a central repository for all documents associated with the programme. This library is essential to the sharing of research and confirmatory analysis. It has facilitated access to information and the development of credible common positions.

"It was recognised that information sharing may give rise to concern from licensees and vendors of reactor technologies, as the library can be accessed by many members. Access has been restricted to only MDEP participating regulatory bodies and it is important that this be understood by all members and their associated vendors, plant designers, requesting parties and licensees. This should avoid situations in which regulators and owners and operator group members are reluctant to provide sensitive documents.

"In addition, it has been observed that the terms and conditions of information sharing are not standard across member countries, and that export licences and commercial issues need to be given due consideration when sharing information with other members."

4.5.3. Memorandum of Cooperation between CNSC and US NRC

On August 15, 2019, CNSC and US NRC signed an MoC [16] to increase collaboration on the technical reviews of advanced reactor and SMR technologies. Information regarding collaboration between CNSC and US NRC under the MoC are available on both CNSC's and US NRC's websites.

4.5.3.1. Governing documents and agreements

The MoU between CNSC and US NRC for the exchange of technical information and cooperation in nuclear safety matters was signed on August 7–9, 2017 [16]. Key aspects under the MoU include exchange of unclassified safety related information from the other participant on any matter related to the peaceful use of nuclear energy within the other participant's jurisdiction. Details of areas of information exchange and cooperation are provided in the MoU.

The MoU covers the exchange of controlled information between CNSC and US NRC, except for nuclear information related to proliferation-sensitive technologies (e.g. export controlled) and classified information, subject to the requirements of each participant's national laws, regulations and policies and the need to protect the controlled information. The MoU also provides procedures for marking and dissemination of controlled information.

In addition, The MoC between CNSC and US NRC was signed on August 15, 2019 [16]. On March 12, 2024, CNSC, ONR and US NRC signed a new trilateral MoC which supersedes the bilateral MoCs (between CNSC and US NRC, CNSC and ONR, and ONR and US NRC) and further expands their partnership in accordance with the provisions of the respective MoUs pertaining to efforts related to advanced reactor and SMR technologies. The key aspects under the MoC include:

"To further strengthen the USNRC, CNSC, and ONR commitment to share best practices and experience reviewing advanced reactor and SMR technology designs. This may include cooperation in the following areas:

1. Development of shared advanced reactor and SMR technical review approaches that facilitate resolution of common technical questions to facilitate regulatory reviews that address each Participant's national regulations;

2. Collaboration on pre-application activities to ensure mutual preparedness to efficiently review advanced reactor and SMR designs; and

3. Collaboration on research, training, and in the development of regulatory approaches to address unique and novel technical considerations for ensuring the safety of advanced reactors and SMRs."

The ToR between CNSC and US NRC were updated in March 2024 [16], and apply, but are not limited to, joint activities on the following:

"• Nuclear reactors that produce energy

• Water-cooled or non-water cooled (i.e., with alternative coolant technologies) reactors

• Development of shared advanced reactor and SMR technical review approaches that facilitate resolution of common technical questions to facilitate regulatory reviews that address each country's national regulations

• Collaboration on pre-application or pre-licensing activities e.g., design assessment to ensure mutual preparedness to efficiently review advanced reactor and SMR designs

• Collaboration on technical reviews for licensing based on the regulatory framework in each country.

• Collaboration on research, training, and in the development of regulatory approaches to address unique and novel technical considerations for ensuring the safety of advanced reactors and SMRs.

• Exchange of information on advanced reactor and SMR technologies."

Potential outcomes of this collaboration could encompass, but are not restricted to:

"• Collaboration on technical reviews leading to:

- Identification and documentation of specific areas of common focus and agreement on acceptability in view of respective safety frameworks

- Documentation of specific elements of the design where either regulator identified safety concerns.

• Exploration of opportunities for joint research in areas where further research and development could benefit the regulators.

• Collaboration in oversight of research or testing activities related to ongoing or expected regulatory reviews and/or activities intended to be submitted to the regulators for review.

• Joint participation in inspections, especially vendor inspections, related to ongoing or expected regulatory review activities intended to be submitted to both regulators for review.

• Sharing independent regulatory review results, seeking to identify possible areas of potential alignment in regulatory approaches and framework."

Moreover, CNSC and US NRC signed a charter in September 2022 documenting collaboration on a new project under the MoC covering their interest in the BWRX-300 SMR design [17]. The Tennessee Valley Authority and Ontario Power Generation are working together on the industry side to share experience and enhance design standardization. Organizations under this charter include:

- Ontario Power Generation;

- Tennessee Valley Authority;

- General Electric-Hitachi;

- US NRC;

- CNSC.

Workplans under the charter include:

- Advanced Construction Techniques (Steel Composite Diaphragm Plates): The work involves a collaborative review of information on the BWRX-300 Containment and Reactor Building Structural Design (Advanced Construction Technique – Steel Composite Diaphragm Plates). Exchange of information between CNSC and US NRC covers safety review methodologies, regulatory approaches, and treatment of unique aspects of the BWRX-300 Advanced Construction Technique.

- Safety Strategy: The work involves a collaborative review of information on the BWRX-300 Safety Strategy. Exchange of information between CNSC and US NRC will cover regulatory approaches and the application of the IAEA defence in depth methodology.

- Fuel Verification and Validation: The work involves a collaborative review of submissions from General Electric-Hitachi intended to demonstrate that the fuel product selected for the FOAK BWRX-300 SMR is qualified for deployment. Specifically, CNSC will leverage previous US NRC reviews of the GNF2 fuel product in CNSC's review of OPG's construction licence application. Furthermore, CNSC and US NRC will collaborate on aspects specifically related to the BWRX-300 SMR design.

These cooperative activities focus on:

- Alignment on key technical areas that will summarize the findings of CNSC and US NRC collaborative work and identify items of mutual understanding/agreement that could be used in each regulatory body's review process;

- Areas of regulatory alignment;

- Key differences in methodologies;

- Lessons learned from this cooperative initiative and areas for improvement to inform future cooperative work.

4.5.3.2. Collaboration publications and outcomes

Increased collaboration between CNSC and US NRC under the MoC has led to the establishment of several joint reports. These reports document the results of collaborative activities between CNSC and US NRC on advanced reactor and SMR technologies. These reports cover topics that are generally applicable to advanced reactor developers and designers, as well as targeted reports addressing specific technical aspects for individual vendors or potential applicants.

4.5.3.3. Information shared

To date information available to the public and proprietary information has been shared between US NRC and CNSC. Security and safeguards related information have not been shared to date, although that may occur for future collaborative activities.

Controlled information, including proprietary information, is handled under Section III, Use of Information and Allocation of Intellectual Property Rights, of the MoU [16], and is protected from public disclosure under the national laws, regulations, or policies of the country of the participant providing the information.

4.5.3.4. Repositories

CNSC and US NRC use a commercial cloud-based content management, collaboration and file sharing tool, to share information needed for their work under the MoC and to store draft and edit joint reports. This electronic repository is password protected and restricted to those who are currently participating in the MoC activities. The repository is managed by US NRC. In accordance with the agreed protocol in the ToR, US NRC controls permission to access the information in the cloud-based electronic repository.

Interim and Final Joint Reports are made publicly available on both CNSC's and US NRC's websites. Other than the public websites, there is currently no long term document library for the MoC. The cloud-based electronic repository is only used as a temporary repository to exchange information and edit joint reports. Once the joint reports are finalized, they are removed from the cloud-based electronic repository. In addition, a DIO provides the information to each regulatory body and the information protection processes established by the regulatory bodies follows national policies and practices.

4.5.3.5. Lessons learned

The following are lessons learned in the collaboration between CNSC and US NRC:

- To limit access to information to those directly working on the project, only a few CNSC staff have access to General Electric-Hitachi documents in the information portal.

- The use of the cloud-based electronic repository has facilitated information sharing.

- Differences in the criteria for withholding information in each country can make it challenging to mark joint documents that will be publicly available. The project leads in US NRC and CNSC coordinate early with the vendor to get agreement on what information will be withheld and ensure that requirements in both countries are satisfied.

- Early in the collaboration under the MoC, the regulatory bodies' ability to perform collaborative reviews of documents was limited because, while the cloud-based electronic repository was a good tool for sharing information securely, it did not support collaborative work on documents. Changes to the cloud-based electronic repository were implemented to allow US NRC and CNSC staff to edit documents and that has facilitated greater efficiency in developing joint products.

- Although the MoU between US NRC and CNSC allows for the sharing of controlled information when needed, the DIO is generally asked to provide documents containing proprietary information directly to both agencies.

4.5.4. ASN/STUK/SÚJB collaboration

From June 2022 to June 2023, regulatory bodies from France, Finland and the Czech Republic conducted a collaborative review referred to as the Joint Early Review (JER) of the light water SMR, NUWARD SMR, developed in part by Électricité de France S.A. (EDF). The regulatory bodies were the Autorité de Sûreté Nucléaire (ASN), STUK, and the State Office for Nuclear Safety (SÚJB), respectively. In addition, they were joined by the TSOs for ASN and the Institute for Radiological Protection and Nuclear Safety (IRSN) (ASN and IRSN are now merged as ASNR), and for SÚJB, National Radiation Protection Institute (SÚRO). The JER was conducted because of interest expressed by energy companies from the three countries for the future construction of the NUWARD SMR.

The regulatory bodies involved in this initiative consider that this form of collaboration can provide particular benefits to all participating parties. Indeed, for the regulatory bodies, this initiative provided room for sharing of knowledge, experience and detailed national practices on topics which are important for safety, and which are crucial in the licensing process. It also enabled regulatory bodies to acquaint themselves with the SMR design, and thus to anticipate the main regulatory and technical challenges. It enabled EDF to receive timely feedback from the regulatory bodies on topics of the highest importance for its design, when modifications are still relatively easy to make and with an objective to develop a standardized design with a level of safety that is consistent with regulatory bodies' expectations, and thus more likely to be accepted by several countries in the future. This exercise also showed that differences in regulatory frameworks do not necessarily need to be addressed through design changes.

As a result, on September 26, 2023, the following public reports were released regarding NUWARD SMR JER:

- JER - Pilot Phase Closure Report [18], issued by ASN/STUK/SÚJB regulatory bodies;
- JER - Summary Report [19], issued by NUWARD SAS (subsidiary of EDF dedicated to the development of its SMR).

These two public reports are complementary as they present the two sides of the same JER experience and, as such, they offer insights to the whole nuclear community into what was done, how it was done and what was collectively learnt. Considering the success of the pilot phase, three additional regulatory bodies volunteered to join the initiative, and they participated in the second phase of JER.

4.5.4.1. Governing documents and agreements

On February 16, 2022, EDF, STUK, SÚJB and SÚRO signed an NDA to allow EDF, in coordination with ASN and IRSN (now merged into ASNR), to provide STUK, SÚJB and SÚRO with information on the NUWARD SMR technology and safety approach in advance of any future NUWARD SMR licensing activities and to permit STUK, SÚJB and SÚRO to exchange information in relation to common regulatory basis and references as well as potential differences that may have to be addressed in advance of the licensing of NUWARD SMR in Finland and in the Czech Republic. This NDA will be updated to integrate the three additional regulatory bodies which will join the regulatory body's group.

The pilot phase of the JER was ruled by a ToR, signed in April 2022 by the participating regulatory bodies ASN, STUK and SÚJB. This ToR will be updated to integrate the three additional regulatory bodies that will join the existing regulatory bodies group. Moreover, this new version of ToR will rule the implementation of the second phase of JER.

In addition to the ToR, a cooperation framework (mandate) has been established to describe the process of work and the scope of discussion. The mandate for the second phase of JER will be updated and will detail the new scope of discussion such as the following pilot phase topics that are to be developed further:

- Definitions of design basis accidents and design extension conditions;
- List of design extension conditions without significant fuel degradation or core melt and study rules;

- Single failure criterion;

- Loss of offsite power combination in transient studies;

- Pool cooling strategies and technical acceptance criteria;

- Classification of structures, systems and components (SSCs).

And the new topics to discuss, which are:

- Management strategies for design extension conditions;

- Containment and assessment of radiological consequences;

- Criticality risk management;

- Architecture of electrical and instrumentation and controls systems.

4.5.4.2. Collaboration publications and outcomes

For each selected topic, a joint synthesis was prepared by the working group, highlighting common and individual conclusions (such as main convergence and divergence points); these documents were shared with EDF. The regulatory bodies involved in this initiative consider that this form of collaboration can provide particular benefits to all participating parties. Indeed, for the regulatory bodies, this initiative provided room for sharing of knowledge, experience and detailed national practices on topics which are important for safety, and which are crucial in the licensing process.

Through the joint syntheses, the working group addressed their conclusions on the NUWARD SMR. These syntheses also compare the Czech, Finnish and French regulatory frameworks, guides and practices between each other and with the NUWARD SMR design and safety approach.

In addition, the closure report [18], developed by the participating regulatory bodies, provides details of the initiatives and the lessons learned from the involved regulatory bodies and their technical support organizations. The report provides national background information from each participating country to clarify their interest in such initiatives. Then, the objectives and working methods of the JER are described. In the report, the regulatory bodies also share their feedback and lessons learned on the initiative. Finally, the main common and individual views on the different topics of the programme of work are also shared in the report. This represents a high level summary of the synthesis reports.

Moreover, the summary report [19] presents the NUWARD SAS team's experience of the JER. As such, it is meant to be read in conjunction with the combined JER Pilot Phase Closure Report, issued at the same time. The two reports are complementary as they present the two sides of the same JER experience and, as such, they offer insights to the whole nuclear community into what was done, how it was done and what was collectively learnt. The summary report presents the main objectives of the JER, which are:

- Foster exchanges with several European regulatory bodies on the NUWARD SMR design and its safety approach.

- Identify key enablers and conditions to meet licensing expectations in these countries.

- Enable all participants to increase their knowledge of regulatory practices in other jurisdictions.

4.5.4.3. Information shared

The information shared by EDF with the regulatory bodies included both publicly available and proprietary information. No export-controlled information was shared. This information served as the basis for the exchanges and for the production of the reports mentioned in the previous sub-section. It mainly consisted of:

- Extracts from the Safety Options File;

- NUWARD presentations at technical meetings;

- Answers to questions and questionnaires sent out and formalized in meeting minutes.

4.5.4.4. Repositories

No specific document repositories were used by the participating regulatory bodies of the JER. Documents were shared by email. The closure report of the pilot phase issued by participating regulatory bodies was made available on ASN's and STUK's websites. EDF has also released its own summary report on NUWARD's website.

4.5.4.5. Lessons learned

Questions related to the access and sharing of information are best addressed before the beginning of the initiative. This includes the outcomes and deliverables of the initiative, as the regulatory bodies may be willing to share them with the public and the international community.

Mainly due to security concerns, no specific digital tools were setup to support file sharing. Documents were shared by emails. Members agreed that for larger working groups, the use of secure online platforms to share documents could be a way to improve efficiency.

In addition, the following lessons learned regarding information sharing between regulatory bodies and between EDF and the regulatory bodies were identified:

- In order for the information sharing to be successful, the terms of reference of the initiative, agreed by all participants, needs to define the objectives of the cooperation and the working methodology. It needs to be sufficiently flexible to welcome consensus-based changes and adaptations during the initiative and be sufficiently informative to describe the methodology of work and provide guidance for participants.

- Information sharing between EDF and regulatory bodies is governed by bilateral NDAs. These agreements enable the sharing of technical information necessary to feed the review and discussion.

  Continuous improvement and adjustments of the process to better meet the objectives will be sought through discussions between the working group and EDF, especially during the pilot phase of an initiative. Adjustment is most likely to be needed regarding the level of detail of the review and the detail of the information shared: a balance needs to be sought between an overall review of a topic (which helps to quickly achieve worthwhile conclusions), and a more detailed review by technical experts (which provides more added value). This balance is best defined before the beginning of the process.

4.5.5. Rostechnadzor/regulatory bodies of embarking countries cooperation agreements and MoUs

In accordance with the Decree of the Government of the Russian Federation No 339 of 15 April 2014, the Federal Environmental, Industrial and Nuclear Supervision Service (Rostechnadzor) has been appointed the authorized body in the area of cooperation for development of national regulatory systems in use of atomic energy for civil purposes in embarking countries ordering construction of nuclear installations of Russian Federation design.

4.5.5.1. Governing documents and agreements

The legal basis for cooperation between Rostechnadzor and the regulatory authorities of the States ordering the construction of Russian Federation-designed nuclear installations are inter-agency agreements or MoUs between these regulatory bodies on cooperation in the field of nuclear and radiation safety regulation which are normally signed following:

- Intergovernmental agreements on cooperation in the field of peaceful use of atomic energy;

- Intergovernmental agreements on cooperation in construction of NPPs.

Rostechnadzor has signed cooperation agreements with the following regulatory bodies of the embarking countries:

- The Viet Nam Agency for Radiation and Nuclear Safety (under the Ministry of Science and Technology, Viet Nam);

- The Nuclear Regulatory Authority, Türkiye (previously Turkish Atomic Energy Authority);

- The Ministry of Science and Technology, Bangladesh;

- The National Nuclear Regulator, South Africa;

- The Ministry for Emergency Situations, Belarus;

- The Minerals Regulatory Commission, Jordan;

- The Radiation Protection Authority, Zambia;

- The Nuclear Regulatory Authority, Nigeria;

- The State Committee of Industrial Safety, Uzbekistan.

Rostechnadzor has signed the following MoUs with the regulatory bodies of embarking countries:

- The Nuclear and Radiological Regulatory Authority, the Arab Republic of Egypt;

- The State Inspectorate of Environmental and Technical Safety under the Government, the Kyrgyz Republic;

- The Ministry of Health, the Republic of Zambia;

- The Nuclear Energy Regulatory Agency, the Republic of Indonesia;

- The Moroccan Agency for Nuclear and Radiological Safety and Security, the Kingdom of Morocco;

- The Electricity and Nuclear Technology Regulatory Authority, the Plurinational State of Bolivia;

- The Department of Science and Technology - Philippine Nuclear Research Institute, the Republic of the Philippines.

The cooperation agreements and memorandums of understanding are based on the Parties' national legislation and define the scope of bilateral cooperation on safety regulation in the peaceful use of atomic energy in the following main areas:

- Development of regulations and guides in the field of nuclear and radiation safety based on domestic regulatory documents[6];
- Licensing of facilities and activities in the field of peaceful use of atomic energy;
- Oversight and control including the development and implementation of inspection programmes;
- Safety regulation in radioactive waste and spent nuclear fuel management, including transportation and safe storage;
- Supervision of accounting and control of nuclear materials, radioactive substances, radioactive waste as well as supervision of physical protection of nuclear installations, radioactive sources, storage facilities, nuclear materials and radioactive substances;
- Quality control of nuclear installation equipment important to safety;
- Emergency preparedness and response;
- Training of personnel;
- Other areas agreed by the Parties.

The Parties under the cooperation agreements and MoUs may interact in the following ways which are specified and agreed by the Parties in each specific case:

- Exchange of information and documentation;
- Technical visits of experts to attend joint seminars, meetings and consultations;
- Scientific visits, training courses and workshops;
- Implementation of joint projects;
- Other forms agreed by the Parties.

An Administrator is designated by each party who is responsible for the identification of the specific areas of cooperation and for the supervision of the implementation of joint activities under the cooperation agreements or MoUs.

4.5.5.2. Collaboration publications and outcomes

Examples include several activities to support regulatory bodies in embarking countries to review the safety of nuclear installations being constructed with technology from the Russian Federation. For example, a review of safety analysis of a nuclear installation carried out by Rostechnadzor's TSOs considering Russian Federation's regulations and guides, IAEA safety standards and/or other States' regulations and guides.

---

[6] Existing Rostechnadzor regulatory documents that establish requirements for SMR safety are available at: https://docs.secnrs.ru/catalog/FNP/. These documents are in Russian and, in some instances, in English.

4.5.5.3. Information shared

The specifics of information exchange within the framework of cooperation depend on the type of work to be performed, including the pre-licensing review of design materials, and are determined by a separate procedure or subject to additional agreements. The information and documentation are exchanged through the administrators designated by each party. The following apply to the information sharing:

- The Parties will not exchange the information which is specified as a state secret of the Russian Federation or classified information of the other party.

- Information will be clearly marked as 'Confidential' when transferred under the agreements or MoUs or created from the implementation thereof and stated by the transferring party as confidential.

- The party receiving the information marked as 'Confidential' will protect it at a level equivalent to the level of protection applied by the transferring party to such information. Such information will not be disclosed or transferred to a third party without the written consent of the transferring party.

- The Parties will, as much as possible, limit the number of individuals having access to information which the transferring party regards as confidential.

- Such confidential information will be treated as 'official information' of limited distribution in the Russian Federation and protected in accordance with the Russian Federation legislation.

- Such confidential information will be treated as 'restricted information' in the state of the other party and protected in accordance with the national legislation of this state.

- Information transferred under the agreements or MoUs will be used exclusively in accordance with the provisions of the signed documents.

- The parties adequately and effectively protect intellectual property in accordance with the national legislation and international agreements of the Russian Federation and the state of the other party.

4.5.5.4. Repositories

No dedicated information repositories are created to maintain documents generated as part of the discussed activities.

4.5.5.5. Lessons learned

At this time, there are no publicly available lessons learned generated as part of the discussed activities.

4.5.6. International cooperation between Republic of Korea and the United Arab Emirates

Following ROK's first NPP export to the United Arab Emirates (UAE) in December 2009, the need for regulatory and technical support for the UAE emerged. In particular, the export deal, which includes a design, build and operate agreement for the four ROK-designed APR1400 reactors at Barakah, has paved the way to enhance cooperative activities between the two countries in the field of nuclear safety and regulation. With respect to the non-proliferation areas, the collaboration between UAE and the ROK has helped both countries conduct business efficiently in compliance with international nuclear non-proliferation norms. The ROK has also shared its physical protection and cybersecurity practices with the UAE to ensure the nuclear security of the power plant.

With the Cooperation Agreement between the ROK Government and the UAE Government on the Peaceful Use of Nuclear Power signed in June 2009, a series of arrangements regarding nuclear safety regulation areas were signed in 2010 followed by the administrative agreement between the two countries' regulatory bodies in January 2023.

4.5.6.1. Governing documents and agreements

The signing authority or organizations concerned are entitled to govern the relevant documents and agreements for further follow-ups. Table 3 shows the Agreement, Arrangements and implementing arrangements between the ROK and UAE.

TABLE 3. GOVERNING DOCUMENTS AND AGREEMENTS BETWEEN THE ROK AND THE UAE

Agreement Category	Entity from ROK	Entity from UAE	Agreement Purpose	Effective Date
Agreement	Government of the Republic of Korea	Government of the United Arab Emirates	Cooperation in the Peaceful Uses of Nuclear Energy	Signed on June 22nd 2009 Effective on January 12th 2010
Arrangement	The Ministry of Education, Science and Technology (MEST)	The Federal Authority for Nuclear Regulation (FANR)	Exchange technical information, nuclear safety regulation, radiation protection, security, physical protection and export control	May 25th 2010
Implementing Arrangement	The Korea Institute of Nuclear Safety (KINS)	FANR	The exchange of technical information and cooperation in nuclear safety and radiation protection matters	May 25th 2010

TABLE 3. GOVERNING DOCUMENTS AND AGREEMENTS BETWEEN THE ROK AND THE UAE (cont.)

Agreement Category	Entity from ROK	Entity from UAE	Agreement Purpose	Effective Date
Special Arrangement	KINS	FANR	The exchange of technical information and cooperation in nuclear safety and radiation protection matters	22 Jul 2010; (revised on 8 Mar 2011 and 14 Nov 2018)
MoU	KINS	Khalifa University of Science, Technology and Research	The exchange of Information, joint-research and education and training in nuclear safety	18 Dec 2011
Administrative Arrangement	NSSC	FANR	To stipulate the obligations related to nuclear security, safeguards and export controls	15 Jan 2023

In May 2010, the Arrangement was signed between the MEST of the ROK and the UAE's FANR, which outlined bilateral cooperation in nuclear safety, safeguards, physical protection and import and export controls. This Arrangement included provisions for information exchange, joint research, education and training programmes and annual meetings to share progress and discuss issues related to these areas of work.

The two parties agreed to set up a close cooperative relationship in nuclear safety, safeguards, physical protection and import and export controls, including provisions for information exchange, joint research, education and training programmes, and annual meetings to share progress and discuss issues related to these areas of work.

The arrangement between the ROK MEST (currently under NSSC) and UAE's FANR focuses on an exchange of technical information, nuclear safety regulation, radiation protection, security, physical protection and export control. The following are included in the agreement:

- Exchange of information on nuclear safety and radiation protection legislation, regulations and guidelines;

- Exchange of information on major nuclear safety and radiation protection reviews, safety licensing decisions;

- Exchange of information on reports on incidents and other construction and operation experience of major safety or radiation protection significance;

- Exchange or loan of samples, materials, equipment and components;

- Exchange of information on nuclear safeguards, physical protection, export control, and related legislation and current practices;

- Joint research and development in mutually determined areas;

- Education and training programmes;

- Major public information activities.

There is also a Special Arrangement between KINS and UAE's FANR regarding technical information exchange on safety regulations and cooperation of nuclear safety and radiation protection. The objectives of the Special Arrangement include:

- To exchange technical information on nuclear and radiation safety, radioactive waste management, training and human resource development, quality assurance management, knowledge and information management, public communication, nuclear safety research, and licensing administration and management;

- To carry out the cooperation activities on operating experiences of nuclear safety and radiation protection;

- To provide education and training support;

- To implement joint programmes and projects related to safety research, including the use of test facilities and computer programs;

- To support FANR requests for information on nuclear regulatory activities.

The Special Arrangement between KINS and FANR was enacted on July 22, 2011, and revised and given its present name on November 14, 2018, regarding direct support for regulatory body technology. The objectives of the arrangement include:

- To dispatch the ROK's regulatory experts to FANR;

- To provide education and training support for FANR regulatory bodies and English versions of documents related to regulations for facilities such as Shin-Kori Units 3 and 4;

- To offer support of safety review and inspection for each stage of related construction and regulatory procedures;

- To hold KINS-FANR cooperation project review meetings, technical meetings and workshops on current technical issues, etc.;

- To share operating information and regulation experiences on the APR 1400.

Under the revised version of the above-mentioned arrangement, the specific term has changed from Special Agreement to Special Arrangement. In addition, the project review meeting held on semi-annual basis has changed to a biannual occurrence.

The Memorandum of Understanding for the exchange of information, joint-research and education and training in nuclear safety between KINS and Khalifa University of Science, Technology and Research of the United Arab Emirates covers the following areas:

- Publications and information on nuclear safety and radiation protection;

- Nuclear safety and radiation protection related research and development;

- Joint research;

- Related education and training.

Lastly, the Administrative Arrangement between NSSC and FANR was established to stipulate the obligations related to the nuclear security, safeguards and export control with the effect of simplifying procedures of nuclear export permits.

4.5.6.2. Collaboration publications and outcomes

Specific products were produced considering the specific needs of UAE and they vary in accordance with the types, level of depth and breadth of respective agreements, arrangements and MoUs. For example, documents provided by KINS include:

- Reports on nuclear safety and radiation protection legislation, regulations and guidelines; which explain the ROK's regulations and provide translations of relevant material:
  - Atomic Energy Act, Enforcement Decree of the Atomic Energy Act, Enforcement Regulation of the Atomic Energy Act, Regulation on Technical Standards for Nuclear Reactor Facilities and Radiation Safety Control.
  - Guidelines on the Pre-Operation (Facility, Performance) Inspection of Light Water Reactors (LWRs).

- Reports on incidents and other construction and operation experience of major safety or radiation protection significance from experience with the construction and operation of the APR1400 in the ROK:
  - The English version of the APR1400 Standard Design Certificate review report;
  - The English versions of approximately 670 RAI open items and 2100 RAIs for Shin-Kori units 3 and 4 construction permit review;
  - The English version associated with the operating licence documentation including two packages of the RAIs raised during the Shin-Kori units 3 and 4 operating licence review;
  - The English version of The Report of Special Safety Review for ROK NPPs in Operation and Research Reactors conducted with respect to the accident at the Fukushima Daiichi NPP;
  - Evaluation Report on the Application for a Design Approval for the LWR Fresh Fuel Shipping Container.

In March 2011, the Korea Institute of Nuclear Non-proliferation and Control (KINAC) signed an MoU with FANR to establish continuous collaboration between the two organizations. Since then, KINAC has been providing guidance to the UAE on the establishment of an effective nuclear control system by participating in and supporting bilateral meetings organized by the NSSC (see Fig. 3).

In addition, as a part of implementing the action items agreed by two parties, various types of meeting such as bilateral meetings and technical meetings have been producing the relevant documents and reports. The two paragraphs below provide details of meetings between the UAE and the ROK.

The Nuclear Safety and Security Commission and FANR Bilateral Meeting (formerly Annual Meeting) discusses issues related to nuclear control and safety regulation. Initially, the first meeting held in November 2011 was called the Annual Meeting on Export Control. However, starting from the fifth meeting in November 2015, the chief representatives from both sides were promoted to the director general level. In addition, as the scope of cooperation between the two countries expanded to include various fields of nuclear regulation such as sharing information on international transportation of nuclear materials, introducing ROK's experience in physical protection inspection advisory services and discussing education and training cooperation, the name of the meeting was changed to Annual Meeting from the sixth meeting held in November 2016. Following this, during the second High Level Meeting in November 2019 and the ninth Annual Meeting in the same month, it was agreed to broaden the scope of discussion from nuclear control to safety regulation and to raise the profile of the meeting by nominating the lead representative from the ROK to be the NSSC's Chairperson rather than its Director General. A total of 10 bilateral meetings were held through 2022.

The NSSC/KINAC and FANR Technical Meeting is a director-level meeting to address practical issues related to nuclear control between the two countries. At the first Annual Meeting in November 2011, it was suggested to establish a regular working-level meeting to delve deeper into technical matters related to nuclear control. As a result, the first Technical Meeting was convened in Daejeon in July 2012. KINAC provided active support to the UAE in effectively implementing export controls and safeguards, addressing any challenges that may arise in the transfer of goods and technology to Barakah NPP. KINAC also shared its experience in implementing physical protection and cybersecurity and introduced the International Nuclear Nonproliferation and Security Academy's education and training initiatives to encourage FANR to participate. A total of nine Technical Meetings have been organized through 2022.

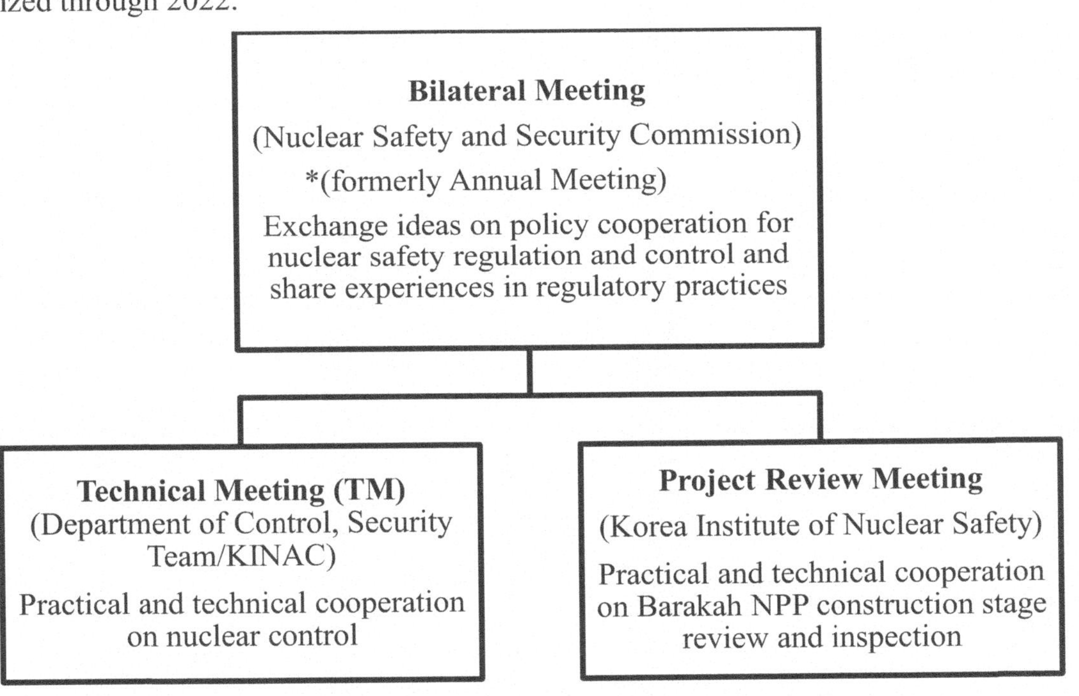

FIG. 3 The ROK-UAE bilateral consultative body.

4.5.6.3. Information shared

In general, the types of information shared between the signatory bodies are:

- Nuclear safety and radiation protection legislation, regulations and guidelines;

- Major nuclear safety and radiation protection reviews, safety licensing decisions (including design information/controlled information) - this was translated and transferred by the vendor;

- Nuclear safeguards, physical protection, export control, and related legislation and current practices;

- Joint research and development in mutually determined areas;

- Education and training programmes;

- Major public information activities;

- Nuclear and radiation safety, radioactive waste management, training and human resource development, quality assurance management, knowledge and information management, public communication, nuclear safety research and licensing administration and management.

In terms of proprietary information, the design information was shared directly by the vendor; KINS also provided some design information if necessary and this was allowed as UAE already had the design information from the vendor.

4.5.6.4. Repositories

There are no formally established and publicly available repositories between the UAE and the ROK. Most of the documents between the ROK and the UAE are classed as confidential. In particular, products derived from the follow-up activities of the Agreement, Arrangements and MoU are mostly unavailable to the public being only accessible for those participating in the cooperation.

In the case of agreements between countries, some documents can be accessed, in Korean, on the website of the Ministry of Foreign Affairs[7].

4.5.6.5. Lessons learned

In order to facilitate feedback and to share the relevant information publicly, the products such as documents, reports and meeting minutes may be made available on an open-access website. In addition, it is suggested that the guidelines of level of categorization of the product be established so that the documents be properly classified and marked appropriately in accordance with the required level of protection, confidentiality and sensitivity. Without specific rules on the release of documents, most of products remain confidential to the public.

4.5.7. EU SMR partnership collaboration

The initiative emerged following the first European Union (EU) workshop on SMRs in June 2021 in response to the growing interest of the EU member states in this technology.

A major outcome of this workshop was the endorsement of a vision paper, including a proposal to establish a European SMR Partnership, hereafter referred to as Partnership, in the form of a collaboration scheme involving industrial, research and technological organizations, interested customers (i.e. utilities) as well as European policymakers and regulatory bodies. The effort aims to facilitate the start-up of the first European SMRs in the next decade while, at the same time, lessening Europe's strong dependence on foreign industries.

This Partnership focuses primarily on SMR technologies that will be available by 2030, so as to play a significant role in reaching the Net Zero goal by 2050 in Europe. Advanced modular reactors based on new technologies (e.g. technologies reducing significantly the production of radioactive waste or technologies providing heat sources directly usable for different industrial purposes) will also be dealt with within the framework of this Partnership.

---

[7] https://www.mofa.go.kr/eng/index.do

The launch of the Partnership has been preceded by the creation of a Steering Committee during a so-called pre-Partnership phase which was tasked with developing and giving general direction in drafting and rolling out a roadmap that was to be shared and endorsed by all the relevant parties. The Steering Committee has been supported by five work streams essential for SMR technology outlook and safe deployment in the EU:

1. Market integration and deployment;

2. Licensing;

3. Financing and partnership;

4. Supply chain adaptation and innovation;

5. Research and development.

The regulatory tasks related to nuclear safety were carried out under Work Stream 2 which was led by the European Nuclear Safety Regulators Group (ENSREG), an independent expert advisory group to the European Commission composed of senior officials from national regulatory authorities and the Commission. Work Stream 2 focused on possible improvements in pre-licensing safety reviews through increased inter-regulatory body collaboration; it also identified the elements for establishing a European pre-licensing process based on commonly accepted safety reviews from different ENSREG members interested in the licensing of the same SMR design.

Beyond the scope of work of each work stream, exchanges took place between the work streams so as to reach a better common understanding of the issues at stake. The reports of the pre-Partnership were published at the end of June 2023. A Stakeholders Forum was organized in October 2023 to collect the feedback from interested parties who were not directly involved in the work streams and to inform them of the progress made. The pre-Partnership phase was followed by a launch of the European SMR Initiative in early 2024. The overall safety objective of the pre-Partnership has been to identify enabling conditions and constraints towards safe design, construction and operation of SMRs in Europe in the next decade and beyond in compliance with the EU legislative framework in general and to the Euratom legislative framework in particular.

4.5.7.1. Governing documents and agreements

The pre-Partnership phase and the work conducted within Work Stream 2 on licensing were not governed by any formal written documents (MoU, MoC) or agreements (NDAs, etc.). A Steering Committee at the framework level of the pre-Partnership was created to give general direction in drafting and rolling out a roadmap for the SMRs development and deployment in Europe.

4.5.7.2. Collaboration publications and outcomes

For the pre-Partnership phase, the main deliverable of Work Stream 2 is the Work Stream 2 Report [20] on licensing in which the following main outcomes are formulated:

- In several countries, a decision in principle is required to initiate a nuclear project, followed by a licensing process, while in others, a pre-licensing step is optional upon applicant request. Typically, only the future licensee can apply for pre-licensing, allowing regulatory bodies to address safety concerns and provide feedback for better predictability in construction licensing.

- Recognizing the importance of early regulatory assessment for new designs, cooperation among regulatory bodies for pre-assessments of SMRs can identify regulatory issues early and facilitate cross-border licensing predictability. Collaborative safety assessments among interested States are suggested, with outcomes aiding national licensing processes.

- Internationally, various initiatives address SMR challenges, but there is a need to focus on actual cases within the European Partnership Initiative. Coordination among EU countries within the IAEA's Nuclear Power Infrastructure Programme is advised to align outputs with European approaches.

- Codes and standards play a crucial role in proving safety system reliability in nuclear installations. While regulatory bodies often do not mandate specific codes and standards, they expect comprehensive justification from licensees for chosen standards. Designers will detail differences from national standards, ensuring compliance with national regulations. Jointly with other work streams, regulatory bodies expressed flexibility in accepting proposed codes and standards packages but require designers to justify deviations and ensure compliance with all relevant regulations, including those beyond nuclear-specific standards.

The Partnership phase includes the European SMR Industrial Alliance. As several EU member states have already decided to support SMR development in their own State, and based on the outcomes of the pre-Partnership phase, the EU has set up the SMR Industrial Alliance[8]. The goal of the Industrial Alliance is to enhance the sharing of SMR developments between similar technologies of reactors to make the best use of this national support, covering topics such as: (a) sharing common detailed licensing costs; (b) preliminary supply chain investments for long lead equipment; (c) training of necessary human resources; and (d) pressurized water reactor fuel adaptation and manufacturing. The European SMR Industrial Alliance was launched in a format comprising Track 1: 7 working groups, Steering Committee and an Advisory Board; and Track 2: concrete design- (or technology-) related projects serving as test cases to be run under strict commercial protection/non-disclosure agreements.

The involvement of regulatory bodies from EU member states will be represented by ENSREG through an interaction mechanism with the Industrial Alliance focusing on sharing the learning of European regulatory bodies in joint reviews with all EU regulatory bodies. The proposal also includes the continuation of Work Stream 2 of the pre-Partnership within ENSREG allowing for direct interaction with the Industrial Alliance working groups. Regulatory bodies from the EU will be able to participate, either individually or jointly, in Track 2 of the Industrial Alliance which will examine planned projects using specific designs or technologies.

4.5.7.3. Information shared

No controlled information was shared during the pre-Partnership phase. The types of information and the scope of its availability (public or specific types of controlled/restricted information) to be shared at the partnership phase within the EU SMR collaborative initiatives on the regulatory pre-licensing and licensing activities is to be defined.

---

[8] https://single-market-economy.ec.europa.eu/industry/strategy/industrial-alliances/european-industrial-alliance-small-modular-reactors_en#documents

4.5.7.4. Repositories

The licensing work stream used a platform to share information and facilitate joint work – the same platform as ENSREG. All reports from the pre-Partnership phase were published in July 2023 and are available on the nucleareurope website[9].

4.5.7.5. Lessons learned

At this time, there are no publicly available lessons learned generated as part of the discussed activities.

---

[9] https://www.nucleareurope.eu/project/european-smr-pre-partnership/

REFERENCES

[1] INTERNATIONAL ATOMIC ENERGY AGENCY, Advances in Small Modular Reactor Technology Developments – A Supplement to: IAEA Advanced Reactors Information System (ARIS) 2022 Edition, IAEA, Vienna (2022).

[2] INTERNATIONAL ATOMIC ENERGY AGENCY, Collaborative Reviews and Effective Leveraging, IAEA Nuclear Harmonization and Standardization Initiative, IAEA-TECDOC-2098, IAEA, Vienna (2025),
https://doi.org/10.61092/iaea.psrj-fkjj

[3] INTERNATIONAL ATOMIC ENERGY AGENCY, Multinational Pre-licensing Joint Regulatory Review: IAEA Nuclear Harmonization and Standardization Initiative, [NHSI Working Group 2 report], IAEA TECDOC (in preparation).

[4] EUROPEAN ATOMIC ENERGY COMMUNITY, FOOD AND AGRICULTURE ORGANIZATION OF THE UNITED NATIONS, INTERNATIONAL ATOMIC ENERGY AGENCY, INTERNATIONAL LABOUR ORGANIZATION, INTERNATIONAL MARITIME ORGANIZATION, OECD NUCLEAR ENERGY AGENCY, PAN AMERICAN HEALTH ORGANIZATION, UNITED NATIONS ENVIRONMENT PROGRAMME, WORLD HEALTH ORGANIZATION, Fundamental Safety Principles, IAEA Safety Standards Series No. SF-1, IAEA, Vienna (2006),
https://doi.org/10.61092/iaea.hmxn-vw0a

[5] INTERNATIONAL ATOMIC ENERGY AGENCY, Application of a Graded Approach in Regulating Nuclear Installations, IAEA-TECDOC-1980, IAEA, Vienna (2021).

[6] UNITED STATES NUCLEAR REGULATORY COMMISSION, Pre-Application Activities for Advanced Reactors,
https://www.nrc.gov/reactors/new-reactors/advanced/who-were-working-with/pre-application-activities.html

[7] UNITED STATES NUCLEAR REGULATORY COMMISSION, SMR Pre-Application Activities,
https://www.nrc.gov/reactors/new-reactors/smr/licensing-activities/pre-application-activities.html

[8] CANADIAN NUCLEAR SAFETY COMMISSION, Pre-Licensing Review of a Vendor's Reactor Design, REGDOC-3.5.4, CNSC, Ottawa (2018),
https://www.nuclearsafety.gc.ca/eng/acts-and-regulations/regulatory-documents/published/html/regdoc3-5-4/

[9] OFFICE FOR NUCLEAR REGULATION, New Nuclear Power Plants: Generic Design Assessment Guidance for Requesting Parties, ONR, Bootle (2019),
https://www.gov.uk/government/publications/new-nuclear-power-plants-generic-design-assessment-guidance-for-requesting-parties/new-nuclear-power-plants-generic-design-assessment-guidance-for-requesting-parties

[10] OFFICE FOR NUCLEAR REGULATION, Licensing Nuclear Installations, ONR, Bootle (2021),
https://www.onr.org.uk/media/30nh5c0f/licensing-nuclear-installations.pdf

[11] INTERNATIONAL ATOMIC ENERGY AGENCY, Format and Content of the Safety Analysis Report for Nuclear Power Plants, IAEA Safety Standards Series No. SSG-61, IAEA, Vienna (2021).

[12]   INTERNATIONAL ATOMIC ENERGY AGENCY, Functions and Processes of the Regulatory Body for Safety, IAEA Safety Standards Series No. GSG-13, IAEA, Vienna (2018).

[13]   INTERNATIONAL ATOMIC ENERGY AGENCY, Applicability of IAEA Safety Standards to Non-Water Cooled Reactors and Small Modular Reactors, IAEA Safety Reports Series No. 123, IAEA, Vienna (2023).

[14]   INTERNATIONAL ATOMIC ENERGY AGENCY, Security of Nuclear Information, IAEA Nuclear Security Series No. 23-G, IAEA, Vienna (2015).

[15]   OECD NUCLEAR ENERGY AGENCY, Multinational Design Evaluation Programme, Phase 1 Summary Report, 2006–2021, OECD NEA, Paris (2021).

[16]   CANADIAN NUCLEAR SAFETY COMMISSION, UNITED STATES NUCLEAR REGULATORY COMMISSION, Terms of Reference for the Memorandum of Cooperation on Advanced Reactor and Small Modular Reactor Technologies between the Canadian Nuclear Safety Commission and the United States Nuclear Regulatory Commission, CNSC, Ottawa (2020).

[17]   UNITED STATES NUCLEAR REGULATORY COMMISSION, Collaborative Projects with Canada, https://www.nrc.gov/reactors/new-reactors/advanced/who-were-working-with/international-cooperation/nrc-cnsc-moc/collaboration.html

[18]   AUTORITÉ DE SÛRETÉ NUCLÉAIRE (ASN), RADIATION AND NUCLEAR SAFETY AUTHORITY (STUK), STATE OFFICE FOR NUCLEAR SAFETY (SÚJB), NUWARD, NUWARD SMR, Joint Early Review Pilot Phase Closure Report, NUWARD, Paris (2023), https://www.nuward.com/sites/nuward/files/2023-09/NUWARDSMR_JointearlyReview_ASNSUJBSTUK.pdf

[19]   NUWARD EDF GROUP, NUWARD SMR Joint Early Review Summary Report, NUWARD, Paris (2023), https://www.nuward.com/sites/nuward/files/2023-09/JointEarlyReviewNUWARDSMR.pdf

[20]   EUROPEAN NUCLEAR SAFETY REGULATORS GROUP (ENSREG), NUCLEAREUROPE, SUSTAINABLE NUCLEAR ENERGY TECHNOLOGY PLATFORM (SNETP), European SMR pre-Partnership Reports Workstream 2 - Licencing, ENSREG, nucleareurope, SNETP, Brussels (2022).

ANNEX I. SUGGESTED MEMORANDUM OF COOPERATION

This template for a Memorandum of Cooperation is intended to assist States interested in entering into common understanding with other States regarding cooperative and collaborative NHSI activities, including the exchange of information. Its provisions may be adjusted based on user preferences or their respective domestic legal requirements.

MEMORANDUM OF COOPERATION [*UNDERSTANDING*]

In the area of exchanging information associated with pre-licensing and licensing activities for advanced reactors

Between [Regulatory Body A], [Regulatory Body B], and [Regulatory Body C], [etc.] hereinafter referred to as the "Parties", and individually as a "Party"

CONSIDERING the significance of using nuclear energy for peaceful purposes for the benefit of their countries;

CONSIDERING that Parties represent the States that are Members of the International Atomic Energy Agency (hereinafter referred to as IAEA) and are signatories to its various Conventions, such as [Convention on Nuclear Safety, the Joint Convention on the Safety of Spent Fuel Management and on the Safety of Radioactive Waste Management, the Convention on Early Notification of a Nuclear Accident, the Convention on Assistance in the Case of a Nuclear Accident or Radiological Emergency; etc.] *[Note: The list of Conventions may vary depending on the cooperating States and may include references to national level bilateral agreements as well]*;

EMPHASIZING the essential role and efforts of IAEA to improve safety regulation infrastructure in States related to advanced nuclear reactors and the importance of implementation of the Nuclear Harmonization and Standardization Initiative (hereinafter referred to as NHSI);

EMPHASIZING the importance of cooperation and exchange of information in ensuring safe and secure deployment of advanced reactors in each Party's country and worldwide;

EMPHASIZING the necessity to carefully balance the exchange of information with the national legal framework of each Party, in order to preserve sovereignty while promoting collaborative efforts intended exclusively for peaceful purposes;

CONSIDERING that the Parties each have mutual interest and existing legal bases and approvals for cooperation and exchanging information during advanced reactor pre-licensing and licensing reviews;

RECOGNIZING the value of collaboration in nuclear information exchange as pivotal for advancing scientific knowledge, enhancing technical capabilities and upholding the highest standards of nuclear safety and security;

Have reached an understanding on the following:

[Optional: DEFINTIONS

For the purpose of this Memorandum of Cooperation (hereinafter referred to as MoC)

a) The expression "..." means
b) The expression "..." means

(Users may choose to include specific definitions if their respective legal provisions or practice of the Parties require or deem necessary)]

ARTICLE I: SCOPE

1. To the extent they may be permitted to do so under the laws, regulations and policies of their respective countries, and within the limits of available resources and subject to the availability of funds, the Parties intend to cooperate to the maximum extent practicable during their respective pre-licensing and licensing reviews of advanced reactors. To that end, each Party may use the collaborative processes and tools developed by the IAEA NHSI Regulatory Track, including processes for engaging in collaborative reviews, joint reviews and leveraging the reviews performed by other regulatory bodies, subject to further specification by the Parties under any necessary specific implementation arrangements described in Article III.

2. In the course of the cooperative activities described above, the Parties may share documents used or produced in the course of the safety review of an advanced reactor. This may include, but is not limited to, documents containing:
    (a) Information relating to a safety assessment of an advanced reactor;
    (b) Specific information about a Party's regulatory review process, results and review criteria;
    (c) Good practices or other feedback on a Party's experiences from previous regulatory cooperation or collaboration efforts.

[The categories above are intended to describe broadly the types of information that can be expected to be exchanged when utilizing processes established by NHSI. Parties using this MoC may define the scope of their cooperative activities as they see fit. They may choose to include more specific descriptions of the information products that are within the scope of the MoC, for example:

    (a) Design basis information, Safety Analysis Report, Site Evaluation Report, or other reports on subjects such as waste and spent nuclear fuel management, accident management and emergency preparedness, or other related documents; or
    (b) Review findings published in technical reports, summary reports and assessment reports in conformity to extant norms on confidentiality.
    (c) Design-specific joint documents such as technical reports and common positions resulting from previous cooperations on harmonization of regulatory reviews.]

ARTICLE II: PROCEDURE

1. Upon the execution of this MoC, each Party intends to designate a person as point-of contact to coordinate its participation in the overall activities covered by this MoC. The Parties are expected to inform each other as well as IAEA of the contact details of the designated coordinators. Unless otherwise agreed by the Parties, all requests for information exchanges are expected to be made through the designated coordinators.

2. The communication between coordinators may be carried out by postal services or through appropriate means of electronic communication. Communication on matters related to this MoC is expected to be coordinated by Parties, located at the following addresses:

Regulatory Body A	Regulatory Body B	Regulatory Body C
XXX	XXX	XXX

3. Each Party is expected to notify the other Party[-ies], as well as IAEA, of any change of their coordinator and provide the name and contact information of the new coordinator.
4. All correspondence relating to this MoC is expected to be in the [......] language.[Optional, only when the MoC was prepared in two or more language versions or the parties establish a common language to be used for the purposes of cooperation]

ARTICLE III: DISSEMINATION OF INFORMATION

1. Under the MoC each Party may share with the other Party[-ies] without restriction the information that is publicly available or that can be made publicly available. It is understood that each Party retains absolute discretion to determine what information can or cannot be made publicly available on request under its own national framework, and it is expected that before designating information as public, the Party making that designation has carefully considered the nature of the information, any legal restrictions that may apply, and the risks that may arise from its public disclosure.
2. The Party transmitting the information does not provide any warranty as to its suitability for any particular use or purpose, and the Parties intend that reliance upon or conclusions drawn from such information is to be at each Party's own risk and does not give rise to any liability of any other Party.
3. Information exchanged pursuant to this MoC may be used only for the purpose of the cooperative activity and not for any other purpose without the approval of the Party who provided the information.
4. If in furtherance of the cooperative activities described by this MoC the Parties intend to share sensitive or non-public information, the Parties intend to enter into subsequent agreements in accordance with their national laws, regulations and policies.*
5. If a Party determines it is necessary to obtain the consent or authorization of another interested party prior to exchanging information, that Party is expected to make best efforts to obtain that consent or authorization in a timely manner.
6. Nothing in this MoC is intended to transfer any rights, title, or interests in intellectual property from one Party to the other. In the event that activities under this MoC are expected to result, or in fact result, in the creation of intellectual property, the Parties intend to consult with one another, as appropriate and as necessary, on the treatment of such intellectual property and the allocation of rights to such intellectual property, in separate written instruments, consistent with their respective national laws.

ARTICLE IV-FINANCIAL ARRANGEMENTS

1. This MoC defines the basis on which the Parties intend to cooperate and does not constitute a financial obligation to serve as a basis for expenses. Unless otherwise agreed, each Party is expected to cover its own expenses with regard to activities under the scope of this MoC.
2. For the purpose of implementation of this MoC, contracts between organizations authorized by the Parties may be concluded and representatives of other agencies and expert organizations may be involved.

ARTICLE V-DISPUTE SETTLEMENT

1. This MoC represents a mutual understanding and a shared willingness of the Parties; it does not establish any legally binding commitments or enforceable rights or obligations for any of the Parties involved.
2. In the event of a dispute, controversy or claim arising from or in connection with this MoC, including its interpretation, termination, or invalidity, the Parties intend to use their best efforts to amicably settle such dispute through direct negotiation, or some other means mutually agreed upon by the disputing Parties.
3. In case the dispute relates to any of the cooperative tools and processes developed by the NHSI Regulatory Track, the Parties may consult the IAEA and seek for a non-binding guidance in resolving the dispute.

ARTICLE VI-MODIFICATIONS AND AMENDMENTS

1. This MoC may be modified or amended in writing by mutual consent of all Parties.
2. Any modification or amendment enters into effect on the date of its signature by all Parties and becomes an integral part of this MoC.

ARTICLE VII-FINAL PROVISIONS

1. This MoC may supplement any existing cooperative activities that Parties may already be engaging in pursuant to separate written agreements. This MoC does not supplant or supersede any such agreements.
2. This MoC becomes effective upon signature by both/all Parties and is expected to remain in effect for a period of [...] years. Unless otherwise agreed upon, the Parties intend that this MoC will automatically renew for successive periods of [...] years. Any Party may terminate its participation in this MoC at any time by giving to the other Party(s) a written notification to that effect. It is expected that the terminating Party will provide [...] days written notice to the other Parties. [*Include if multilateral*: However, such termination is not intended to affect the continuing validity of the MoC amongst the non-terminating Parties.]
3. This MoC was prepared in the [...] language and in the [Insert any additional translated] languages. In case of inconsistency, the Parties intend that the [...] version prevails.

SIGNED IN [DUPLICATE/...]   AT _________________, on
____

__________________________________   __________________________________
       [Signature]   [Signature]

   (Name, Title, RB "A")   (Name, Title, RB "B")

____

[Signature]

(Name, Title, RB "C")

*As stated in section 4.3.2 of the report, it is presumed that this MoC will be aspirational and non-binding, and that legally binding provisions governing the exchange of controlled information will need to be imposed through subsequent written agreements in accordance with the Parties' domestic legal requirements. Example provisions that may be incorporated into these subsequent written agreements are provided here.

1. Each Party receiving information that has been designated by the transmitting Party as controlled information will respect its confidentiality. For purposes of this Agreement, 'controlled information' refers to any non-public information that is protected from public disclosure under the laws, regulations and policies of the transmitting Party. It includes, but is not limited to, proprietary information (trade secrets, commercial, financial, intellectual property), nuclear security related information, information that is subject to domestic export control laws and regulations, regulatory pre-decisional information not authorized for public release, or any other category of protected information.

2. A Party transmitting controlled information under this Agreement will clearly mark such information on each page of the document at the time of transmittal to ensure the receiving Party is aware of its controlled status. The transmitting Party will mark such information with the following restrictive legend: [Parties to insert agreed-upon restrictive marking, such as 'CONTROLLED' or 'CONFIDENTIAL - DO NOT SHARE WITHOUT THE WRITTEN CONSENT OF (TRANSMITTING PARTY)'. Parties who have national requirements to mark controlled information in a certain manner will include such required markings in addition to this restrictive legend.]

3. The receiving Party will not make public any information designated by the transmitting Party as controlled or otherwise disseminate the information in a manner inconsistent with this Agreement without the prior written consent of the transmitting Party. The receiving Party will limit dissemination of controlled information to its own employees who have a requisite need to know the information in order to perform official duties. The receiving Party may only share the controlled information with individuals from outside organizations, such as contractors, TSOs, or other support organizations, if they are supporting the Party in an activity within the scope of this Agreement; those individuals similarly have a requisite need to know; and their support activities are governed by a confidentiality agreement with terms that are equally or more protective than this Agreement.

4. The Parties agree to protect controlled information they receive pursuant to this Agreement including those transmitted to outside organizations in accordance with para. 3, as follows: [Parties to insert any applicable information security protections required by their respective national legal framework, such as controls relating to storage, handling, or proper destruction of controlled information. Parties also need to ensure that any controls described in this section are also applied to information shared outside the organization as described in para. 3.]

5. Any Party that makes a copy of controlled information or reproduces the controlled information into a derivative document will protect the controlled information to the same extent as the original copy. The Party is also responsible for ensuring the restrictive legend is placed on any copies made or derivative documents that include the controlled information.

6. Parties may only use controlled information received pursuant to this Agreement to further the purpose of the cooperative activity for which the information was sent, and not for any other purpose unless with the express approval of the Party who provided the information.

7. In the event a Party suspects or confirms there has been an unauthorized disclosure of controlled information it received pursuant to this Agreement, that Party will promptly notify the party who transmitted the controlled information of the circumstances of the unauthorized disclosure and its efforts to recover the information or minimize further unauthorized disclosure. This notification is to be provided regardless of fault.

8. If a Party who receives controlled information pursuant to this Agreement receives a request for the disclosure of the information under any applicable national law governing public transparency and disclosure of government records, the receiving Party commits to withholding such information from public disclosure to the maximum extent permitted under its national laws. If a Party is compelled under its national laws or an order of a competent court of jurisdiction or other legal authority to release controlled information pursuant to these laws, the Party being compelled to disclose the information will promptly inform the transmitting Party.

9. In the event a Party withdraws from this Agreement or the Agreement is otherwise terminated, each Party will continue to treat any controlled information received, as well as transmitted to outside organizations, as confidential and will continue to protect the information in accordance with the Agreement's terms. Upon withdrawal or termination of the Agreement, the party who transmitted controlled information will instruct the other Party on whether the information is to be returned, destroyed, or if other action is to be taken.

ANNEX II. OVERVIEW OF MEMBER STATES PRE-LICENSING AND LICENSING PROCESS INFORMATION REQUIREMENTS

This annex provides examples of how regulatory bodies from Member States review advanced reactors with a focus on the information typically considered during: (a) the processes for early regulatory engagement with a reactor vendor; (b) the pre-licensing process; and (c) the licensing process.

Examples for these three categories have been provided for Canada, Indonesia, the Republic of Korea (ROK), Slovakia, Sweden and United States of America (USA), as applicable and available.

II–1. EARLY ENGAGEMENT WITH A REACTOR VENDOR

II–1.1. United States of America

Pre-application in the USA is optional, and applicants may consider the guidance in the following US NRC documents:

- Pre-application Engagement to Optimize Advanced Reactors Application Reviews (draft) [II–1].

- A Regulatory Review Roadmap for Non-Light Water Reactors [II–2].

II–1.2. Indonesia

In accordance with Act No. 10/1997 on Nuclear Energy [II–3], an applicant may contact Indonesia's Nuclear Energy Regulatory Agency for more explanation on the licensing requirements and the associated procedures.

II–2. PRE-LICENSING PROCESS

II–2.1. Canada

The topics included in the vendor's design review [II–4] include:

- General plant description, defence in depth, safety goals and objectives, dose acceptance criteria.

- Classification of structures, systems and components.

- Reactor core design.

- Fuel design and qualification.

- Control system and facilities:

  - Main control systems;
  - Instrumentation and control;
  - Control facilities;
  - Emergency power systems;
  - Means of reactor shutdown.

- Emergency core cooling and emergency heat removal systems.

- Containment and confinement and civil structures important to safety.

- Prevention and mitigation of beyond-design-basis accidents and severe accidents.

- Safety analysis (deterministic safety analysis, probabilistic safety analysis) of internal and external hazards.

- Pressure boundary design.

- Fire protection.

- Radiation protection.

- Out-of-core criticality.

- Robustness, safeguards and security.

- Vendor research and development programme.

- Management system of design process and quality assurance in design and safety analysis.

- Human factors.

- Incorporation of decommissioning in design considerations.

In addition:

- A vendor can initiate a phase 2 review once the design's basic engineering programme is either well under way or completed. The results of a phase 2 review assist the vendor's development of a preliminary safety analysis report, as part of the preparations in support of an applicant for an eventual (site-specific) application for a licence to construct.

- This phase focuses on identifying if any potential fundamental barriers to licensing exist or are emerging with respect to the reactor design. Phase 2 serves to give CNSC a significant level of assurance that the vendor has taken into account CNSC design requirements. Consideration is also given to the extent to which generic or outstanding safety issues have been resolved. In addition, CNSC staff conduct an audit of the design process, to verify that it has been implemented correctly and in accordance with the vendor's policies and procedures.

- For the phase 2 review, particular attention is paid to the review focus areas where there are new design features or approaches used in the design, to ensure that the vendor has performed or planned testing and analysis work to support the adequacy of the design.

II–2.2. United States of America

In the USA, Regulatory Guide 1.206, Applications for Nuclear Power Plants [II–5], provides guidance for the scope and contents of a design certification application, which includes:

- Site parameter envelope;

- Design of structures, systems, components and equipment;

- Reactor internals;

- Reactor coolant and connected systems;

- Engineered safety features;

- Digital instrumentation and controls/electrical power;

- Auxiliary systems;

- Steam and power conversion systems;

- Radioactive waste management and radiation protection;

- Conduct of operations;

- Initial test programme and inspections, tests, analyses and acceptance criteria;

- Transient and accident analysis;

- Technical specifications;

- Quality assurance programme;

- Human factors engineering;

- Severe accidents.

Additional information about design certification can be found in Section II–3.3.

II–2.3. Indonesia

A draft regulation on pre-licensing in Indonesia is being formulated but has not yet been established.

II–2.4. Sweden

Even though there is no pre-licensing process established in Sweden, expectations are that the generic licensing discussed in Swedish Radiation Safety Authority (SSM) steering document STYR2011-131 [II–6] will be applied in a graded approach for pre-licensing. See Section II–3.6.

II–3. LICENSING PROCESS

II–3.1. Canada

A licence to construct enables a licensee to construct, commission and operate some components of the facility (e.g. security systems). Some commissioning activities may be allowed in order to demonstrate the facility has been constructed in accordance with the approved design and that the SSCs important to safety are functioning as intended.

An application for a licence to construct contains more detailed information about the design of the facility and the supporting safety case than would be needed for vendor design review. The applicant is to demonstrate that the proposed design of the facility conforms to regulatory requirements and will provide for the safe operation on the designated site over the proposed lifetime of the facility.

The applicant is expected to address all follow-up activities identified during the environmental assessment, including those relevant to the design, construction and commissioning stages and verify that any outstanding issues from the site preparation stage have been resolved.

The specific information required for an application for a licence to construct a Class I nuclear facility is listed in sections 3 and 5 of the Class I Nuclear Facilities Regulations [II–7]. Examples of information submitted in support of an application to construct are:

- A description of the proposed design of the facility, taking into consideration the physical and environmental characteristics of the site;

- Environmental baseline data about the site and surrounding area;

- A preliminary safety analysis report, showing the adequacy of the design;

- Description of measures to mitigate the effects on the environment and health and safety of persons that may arise from the construction, operation or decommissioning of the facility;

- Information on the potential releases of nuclear material and other hazardous materials, and proposed measures to control them;

- Programmes and schedules for recruiting and training for workers who conduct licensed activities;

- Public information and disclosure programme to keep the public and target audiences informed of the anticipated effects of the facility's construction activities on their health and safety and on the environment;

- Updated preliminary decommissioning plan;

- Proposed financial guarantee for the activities to be licensed under the licence to construct.

The review of the application focuses on determining whether the proposed design, the safety analysis and other required information meet regulatory requirements. The review involves rigorous engineering and scientific analysis, taking into consideration national and international standards and best practices in nuclear facility design and operation. CNSC also verifies that any outstanding issues from the site preparation stage have been resolved. The protective zone set at the time of the licence to prepare site is also confirmed (or revised) during review of the licence to construct.

For the latter part of construction, regulatory attention focuses on the commissioning programme and associated activities, to demonstrate to the extent practicable that all the SSCs have been built and function as intended. CNSC regulatory document REGDOC-1.1.2, Licence Application Guide: Licence to Construct a Reactor Facility [II–8], provides guidance on the information to be submitted for a licence to construct.

A licence to operate will enable a licensee to complete final commissioning activities and to operate the facility. Commissioning activities provide assurance that the facility has been properly designed and constructed and it is ready for safe operation.

The specific information required for an application for a licence to operate a Class I nuclear facility is in sections 3 and 6 of the Class I Nuclear Facilities Regulations [II–7]. Examples of information submitted in support of an application to operate are:

- A description of the structures, systems and equipment of the facility, including their design and operating conditions;

- The final safety analysis report;

- The proposed measures, policies, methods and procedures for:

  o Commissioning systems and equipment;
  o Operating and maintaining the nuclear facility;
  o Handling nuclear substances and hazardous materials;
  o Controlling the release of nuclear material and other hazardous materials into the environment;

- o Preventing and mitigating the effects on the environment and health and safety, which result from the operation and subsequent decommissioning of the facility;
  - o Assisting offsite authorities in emergency preparedness activities, including assistance to deal with an accidental offsite release;
  - o Developing and maintaining nuclear security.
- Public information and disclosure programme to keep the public and target audiences informed of the anticipated effects of the facility's operation on their health and safety and on the environment;
- Updated preliminary decommissioning plan;
- Proposed financial guarantee for the activities to be licensed under the licence to operate.

The first licence to operate the facility is typically issued with conditions (hold points). All the relevant commissioning tests are to be satisfactorily completed before the hold points can be removed. CNSC regulatory document REGDOC-1.1.3, Licence Application Guide: Licence to Operate a Nuclear Power Plant [II–9], provides guidance on the information to be submitted for a licence to construct.

CNSC's regulatory framework includes the application guides covering the following topics:

- Management system;
- Human performance management;
- Operating performance;
- Safety analysis;
- Physical design;
- Fitness for service;
- Radiation protection;
- Conventional health and safety;
- Environmental protection;
- Emergency management and fire protection;
- Waste management;
- Security;
- Safeguards and non-proliferation;
- Packaging and transport;
- Reporting requirements;
- Public and Aboriginal engagement;
- Financial guarantees.

II–3.2. Republic of Korea

The licensing procedures for nuclear installations in ROK consist of two steps: the construction permit and the operating licence, pursuant to the Nuclear Safety Act (NSA) [II–10]. When necessary, the licensee may apply for Standard Design Approval (SDA) and Early Site Approval.

As the design of NPPs with an enhanced level of safety could be standardized depending on demand, a new licensing system, the SDA system, was issued to improve regulatory effectiveness. The SDA system ensures the validation of an approved standard design without imposing additional regulatory requirements during a certain period of time by the law and basically excludes safety reviews for the parts of NPPs that refer to a previously approved standard design.

In order to start limited construction work on a proposed site before the construction permit is issued, an applicant may apply for an Early Site Approval. The applicant submits an application accompanied by a site survey report and a radiological environmental impact assessment report to the Nuclear Safety and Security Commission (NSSC). Based on the results of the safety review by Korea Institute for Nuclear Safety (KINS) for Early Site Approval, the NSSC will grant an approval. The safety review is to evaluate the adequacy of the proposed nuclear site and the radiological impacts on the environment surrounding the nuclear installation.

In addition, the applicant for a construction permit for a nuclear installation submits an application accompanied by a radiological environmental report, a preliminary safety assessment report (PSAR), a construction quality assurance plan and a decommissioning plan to the NSSC. The NSSC issues a construction permit after deliberation of the application documents based on the results of the safety review of the application submitted by KINS.

The safety review of the application for a construction permit is conducted to confirm that the site and the preliminary design of the nuclear installation conform with the relevant regulatory requirements and technical guidelines. It includes safety reviews on the principles and concepts of reactor facility design, the implementation of regulatory criteria in due course, the review of environmental effects resulting from the construction, and a proposal for minimizing those effects.

The radiological environmental impact assessment report submitted for early site approval and construction permit need to include the opinions of local residents (by holding a public hearing, if needed). In accordance with the law, in the course of the construction permit application of Shin-Hanul Units 3 and 4, the opinions of local residents on the radiological environmental report were reflected. The construction permit applicant prepared a draft environmental report and submitted it to the NSSC and relevant local authorities to receive feedback from relevant agencies. Relevant local authorities were notified in major daily magazines that the report was now open to public. The same procedure was applied to hold a public hearing on the report for the local residents. The construction permit applicant reviewed the opinions gathered from the relevant agencies and also from the residents and reflected them in the final radiological environmental impact report.

The criteria for a construction permit for a nuclear installation are specified in the NSA, as follows:

- Securing the technical capability necessary for the construction of nuclear installations;

- Conformance of the location, structures and components of a nuclear installation to the technical standards provided in the NSSC regulations in such a way that there is no impediment to the protection of humans, materials and the public against hazards caused by radioactive materials, etc.;

- Compliance with the criteria set in the Presidential Decree to prevent hazards to public health and the environment due to radioactive materials which may accompany the construction of nuclear installations;

- Compliance of the quality assurance programme with standards specified in the NSSC regulations;

- Compliance of the decommissioning plan with standards specified in the NSSC regulations.

The technical requirements for the location, structure and equipment of reactor facilities are specified in the Regulations on Technical Standards for Nuclear Reactor Facilities, Etc. [II–11]. The more specific regulatory requirements, if necessary, are prescribed in the NSSC Notices.

The applicant for an operating licence for a nuclear installation submits to the NSSC an application accompanied by technical specifications for operation, a final safety assessment report (FSAR), an accident management plan, a quality assurance plan for operation, a radiological environmental report, a decommissioning plan, and a plan to release radioactive material in liquid and gas form. The NSSC will issue an operating licence primarily based on the results of the safety review of the application as well as the results of pre-operational inspections by KINS.

The safety review of the application for an operating licence is conducted to confirm that the final design of the nuclear installation conforms with the relevant regulatory requirements and technical guidelines and that the nuclear installation may continue to operate throughout its lifetime.

For an amendment to the operating licence such as a change in the technical specifications or in the design that affects or might affect the safety of operating nuclear installations, it is necessary to obtain approval from the NSSC. The approval for an amendment to the operating licence follows the same procedures as the application for an operating licence. A safety review is, however, to be conducted that includes all aspects of safety that are actually or potentially affected by the amendment to the operating licence.

The criteria for an operating licence for a nuclear installation are specified in the NSA, as follows:

- Securing the technical capability necessary for the operation of the nuclear power reactors and related facilities;

- Conformance of the performance of the nuclear power reactors and related facilities to the technical requirements, as prescribed by the NSSC regulations, in such a way that there is not any impediment to the protection of humans, materials and the public against hazards caused by the radioactive materials;

- No identified impediments to the protection of the public and the environment against radioactive materials that may accompany the operation of the nuclear reactors and related facilities, in accordance with the Presidential Decree;

- The quality assurance programme, decommissioning plan and accident management plan are to meet the criteria provided in the NSSC regulations.

The technical standards are prescribed in Regulations on Technical Standards for Nuclear Reactor Facilities, Etc. [II–11]; six articles regarding the technical capability on operation; 38 articles regarding the technical standards for the performance of nuclear installations; and 18 articles regarding the technical standards for the quality assurance programme. Regarding reactor operation, detailed regulatory requirements are specified in a NSSC notice.

II–3.3. United States of America

Design certifications and early site permits under Title 10 of the Code of Federal Regulations (CFR) Part 52 are comparable to getting a licence; therefore, even though they occur before the combined licence stage, they are not considered pre-licensing processes. The design certifications include final design level information that is equivalent to what is found in the FSAR of a combined licence. The FSAR of an approved design certification will be incorporated by reference into the FSAR for a combined licence application.

The 10 CFR Part 52.47, Contents of applications; technical information [II–12], may be summarized as follows:

- The application is to contain a level of design information sufficient to enable US NRC to judge the applicant's proposed means of assuring that construction conforms to the design and to reach a final conclusion on all safety questions associated with the design before the certification is granted.

- The information submitted for a design certification is to include performance requirements and design information sufficiently detailed to permit the preparation of acceptance and inspection requirements by US NRC, and procurement specifications and construction and installation specifications by an applicant.

- US NRC will require, before design certification, that information normally contained in certain procurement specifications and construction and installation specifications be completed and available for audit if the information is necessary for US NRC to make its safety determination.

- The application is to contain an FSAR that describes the facility, presents the design bases and the limits on its operation, and presents a safety analysis of the structures, systems and components and of the facility as a whole, and includes the information listed in 28 subparts.

- An application for certification of a nuclear power reactor design that is an evolutionary change from light water reactor designs of plants that have been licensed and in commercial operation before April 18, 1989, is to provide an essentially complete NPP design except for site-specific elements such as the service water intake structure and the ultimate heat sink.

- An application for certification of a nuclear power reactor design that differs significantly from the light water reactor designs described above or uses simplified, inherent, passive, or other innovative means to accomplish its safety functions is to provide an essentially complete nuclear power reactor design except for site-specific elements such as the service water intake structure and the ultimate heat sink, and meets the requirements of 10 CFR 50.43(e) [II–13].

Applications for construction permits under 10 CFR Part 50 in the US typically contain information on a preliminary design. For a high-level description of the US NRC licensing processes in 10 CFR Parts 50 and 52, see Ref, [II–14]. The following references provide details of additional topics associated with licensing in the US:

- Reference [II–5] provides guidance for applications for a light water reactor early site permit, standard design certification and combined licence applications under 10 CFR Part 52.

- Reference [II–15] provides guidance to US NRC staff to facilitate the safety review of light water power reactor construction permit applications under 10 CFR Part 50 and to supplement the guidance in Ref. [II–16].

- US NRC staff issued guidance on the content of final applications for advanced reactors submitted under 10 CFR Parts 50 and 52 associated with the Advanced Reactor Content of Application Project (ARCAP):

  - RG 1.253, Guidance for a Technology-Inclusive Content-of-Application Methodology to Inform the Licensing Basis and Content of Applications for Licenses, Certifications, and Approvals for Non-Light-Water Reactors [II–17];
  - DANU-ISG-2022-01 ARCAP Review of Risk-Informed, Technology-Inclusive Advanced Reactor Applications-Roadmap Interim Staff Guidance [II–18];
  - DANU-ISG-2022-02 ARCAP Chapter 2 Site Information [II–19];
  - DANU-ISG-2022-03 ARCAP Chapter 9 Control of Routine Plant Radioactive Effluents, Plant Contamination and Solid Waste [II–20];
  - DANU-ISG-2022-04 ARCAP Chapter 10 Control of Occupational Dose [II–21];
  - DANU-ISG-2022-05 ARCAP Chapter 11 Organization and Human-System Considerations [II–22];
  - DANU-ISG-2022-06 ARCAP Chapter 12 Post-Construction Inspection, Testing, and Analysis Program [II–23];
  - DANU-ISG-2022-07 ARCAP Risk-informed Inservice Inspection/Inservice Testing Programs for Non-LWRs [II–24];
  - DANU-ISG-2022-08 ARCAP Risk-informed Technical Specifications [II–25];
  - DANU-ISG-2022-09 ARCAP Risk-Informed, Performance-Based Fire Protection Program (for Operations) [II–26].

II–3.4. Slovakia

In Slovakia, there is a written application for approval of the siting of a nuclear installation and a written application for a building permit for construction of a nuclear installation.

Documentation necessary for the written application for approval of the siting of a nuclear installation includes:

- Reference safety report;

- Reference report on the method of decommissioning;

- Project proposal, a document detailing the fundamental physical layout and key technical systems of the nuclear facility, presented at a similar conceptual level as a standard or previously approved design.

- Reference report on the method of radioactive waste management and spent nuclear fuel management;

- Requirements for the quality of the nuclear installation;

- Proposed boundaries of the nuclear installation;

- Proposal for the size of the emergency planning zone for the nuclear installation;

- Assessment of the impact of nuclear installation on the environment, if established by a special regulation, as well as assessment of the potential impact of the surrounding environment on the nuclear installation.

Documentation necessary for written application for building permit for construction of a nuclear installation includes:

- Preliminary safety report proving fulfilment of legal requirements for nuclear safety on the basis of data, which are contemplated in the design;

- Design documentation necessary for the start of construction;

- Preliminary plan for radioactive waste management, spent nuclear fuel management including shipment;

- Preliminary conceptual decommissioning plan;

- Categorization of classified equipment into safety classes;

- Preliminary physical protection plan;

- Quality management system documentation and requirements for quality of the nuclear installation and their assessment;

- Preliminary on-site emergency plan;

- Preliminary limits and conditions for safe operation;

- Preliminary programme of control of nuclear installation prior to its operation;

- Preliminary demarcation of boundaries of the nuclear installation by specifying the data stated in application for approval of the site location;

- Preliminary demarcation of the size of the emergency planning zone of the nuclear installation supported by specific technical data and analyses.

II–3.5. Indonesia

The following items are required for a construction licence (C) and/or operation licence (O):

- Design Information Questionnaire for safeguards (C);

- Safety analysis report or safety case (SAR for C, and FSAR for O);

- Operational limits and conditions or technical specifications (C, O);

- Management system (C, O);

- Radiation protection and safety programme (C, O);

- Safeguards system (C, O);

- Physical protection programme (C, O);

- Ageing management programme (C, O);

- Decommissioning programme (C, O);

- Emergency preparedness and response arrangements (C, O);

- Construction plan (C);

- Environmental licence from the MOE (C):

- Commissioning plan (O);

- Maintenance programme (O);

- Construction activity report (O);

- Technical drawings of the as built nuclear reactor (O).

II–3.6. Sweden

The licensing in Sweden is based on different versions of the SAR as follows:

- Licensing preliminary safety analysis report (LPSAR[10]) is basis of the application for a new facility;

- PSAR is basis of the application for a licence/approval to construct;

- FSAR is basis of the application to commission;

- SAR is basis for operation.

When a licence is granted (based on the LPSAR) a stepwise review is done based on the SAR steps described above. As the licensing and design process gets closer to construction and commissioning, also other type of documentation, including operational limits and conditions (OLC) and a management system for the operating organization need to be prepared in more detail, in order to be in place for the final decision on approval/permit for operation (not a separate licence).

In the Swedish Radiation Safety Authority's (SSM) regulations for NPPs[11], there are also requirements related to the design process that would be important for a design review (some may be integrated in the SAR, but not necessarily all types of information), for example:

- Choice/application of:

  o Design solutions;

  o Standards and guides;

  o Materials;

  o Manufacturing processes;

  o Installation processes;

  o Qualification processes;

---

[10] Licensing preliminary safety analysis report or pre-construction safety analysis report.

[11] The SSM's website with links to regulations for nuclear power is:
https://www.stralsakerhetsmyndigheten.se/publikationer/?publicationTypes=F%c3%b6reskrifter&area=K%c3%a4rnkraft

- o Other relevant factors of relevance for the selected design to meet safety and security requirements.

- Competence and technical expertise included/active during the design process (i.e. all relevant safety and security issues).

- Use of experience of relevant/comparable designs, including operational experience:

  - o Planned activities for verification and validation of designs solutions in order to assess and verify that the design meets applicable safety and security requirements.

- Based on requirements for management of modifications (including introduction of new elements in design, operation or organization) there are requirements on safety demonstration, as a process for review and demonstration to the authority based on information about:

  - o Background, scope, objectives of the proposed design;
  - o Organization, competences and responsibilities for development of the proposed design,;
  - o Identified relevant safety and security issues;
  - o Claims, arguments and evidence that identified safety and security issues are managed and applicable requirements are fulfilled;
  - o Description of related activities and reports on the proposed design (e.g. reviews, plans);
  - o Independent safety review for the proposed design.

- The purpose of the demonstration is for the applicant/licensee to make a comprehensive and logical presentation of all relevant information (the experience from modifications to existing plants is that the documentation and basis for regulatory review may be scattered, not in a clear way building evidence on the fulfilment of relevant regulatory expectations).

- In general, during a review of modifications or licence applications, it is common for SSM to require references etc. based on the applicant's claims to follow a certain standard or method. Bilateral agreements with other authorities give prerequisites for such issues, as well as for follow-up questions or discussions, especially in cases when information is not openly or easily accessible.

REFERENCES FOR ANNEX II

[II–1] UNITED STATES NUCLEAR REGULATORY COMMISSION, DRAFT Pre-application Engagement to Optimize Advanced Reactors Application Reviews, US NRC, Rockville (2021),
https://www.nrc.gov/docs/ML2114/ML21145A106.pdf

[II–2] UNITED STATES NUCLEAR REGULATORY COMMISSION, A Regulatory Review Roadmap for Non-Light Water Reactors, US NRC, Rockville (2017),
https://www.nrc.gov/docs/ML1731/ML17312B567.pdf

[II–3] REPUBLIC OF INDONESIA, Act No. 10 of 1997 on Nuclear Energy, Jakarta (1997),
https://jdih.bapeten.go.id/unggah/dokumen/peraturan/85-full.pdf

[II–4] CANADIAN NUCLEAR SAFETY COMMISSION, REGDOC-3.5.4, Pre-Licensing Review of a Vendor's Reactor Design, CNSC, Ottawa (2018) https://www.nuclearsafety.gc.ca/eng/acts-and-regulations/regulatory-documents/published/html/regdoc3-5-4/

[II–5] UNITED STATES NUCLEAR REGULATORY COMMISSION, Applications for Nuclear Power Plants, Regulatory Guide 1.206, US NRC, Rockville (2018), https://www.nrc.gov/docs/ML1813/ML18131A181.pdf

[II–6] SWEDISH RADIATION SAFETY AUTHORITY, STYR2011-131, Preparation of licences and review of licence conditions concerning nuclear installations and other complex installations where radiation is used, Stockholm (2010).

[II–7] MINISTER OF JUSTICE (CANADA), Class I Nuclear Facilities Regulations, SOR/2000-204, Ottawa (2024), https://laws.justice.gc.ca/PDF/SOR-2000-204.pdf

[II–8] CANADIAN NUCLEAR SAFETY COMMISSION, REGDOC-1.1.2, Licence Application Guide: Licence to Construct a Reactor Facility, Version 2, CNSC, Ottawa (2022), https://www.nuclearsafety.gc.ca/eng/acts-and-regulations/regulatory-documents/published/html/regdoc1-1-2-v2/index.cfm

[II–9] CANADIAN NUCLEAR SAFETY COMMISSION, REGDOC-1.1.3, Licence Application Guide: Licence to Operate a Nuclear Power Plant, Version 1.2, CNSC, Ottawa (2022), https://www.cnsc-ccsn.gc.ca/eng/acts-and-regulations/regulatory-documents/published/html/regdoc1-1-3-v1-2/

[II–10] NUCLEAR SAFETY AND SECURITY COMMISSION, Nuclear Safety Act, Act No. 10911, Seoul (2011), https://elaw.klri.re.kr/eng_service/lawView.do?hseq=45486&lang=ENG

[II–11] NUCLEAR SAFETY AND SECURITY COMMISSION, Regulations on Technical Standards for Nuclear Reactor Facilities, Etc., Seoul (2020), https://www.nssc.go.kr/attach/namo/files/000001/20200131181134400_L50GVEKZ.pdf

[II–12] UNITED STATES NUCLEAR REGULATORY COMMISSION, Title 10 of the Code of Federal Regulations Part 52.47 Contents of applications; technical information, Washington, https://www.ecfr.gov/current/title-10/chapter-I/part-52/subpart-B/section-52.47

[II–13] UNITED STATES NUCLEAR REGULATORY COMMISSION, Title 10 of the Code of Federal Regulations Part 50.43 Additional standards and provisions affecting class 103 licenses and certifications for commercial power, Washington, https://www.ecfr.gov/current/title-10/chapter-I/part-50/subject-group-ECFRdf2af20f8a72d91/section-50.43

[II–14] UNITED STATES NUCLEAR REGULATORY COMMISSION, Nuclear Power Plant Licensing Process, NUREG/BR-0298, Rev. 2, US NRC, Rockville (2004), https://www.nrc.gov/docs/ML0421/ML042120007.pdf

[II–15] UNITED STATES NUCLEAR REGULATORY COMMISSION, Safety Review of Light-Water Reactor Construction Permit Applications, Interim Staff Guidance DNRL-ISG-2022-01, US NRC, Rockville (2022), https://www.nrc.gov/docs/ML2218/ML22189A101.pdf

[II–16] UNITED STATES NUCLEAR REGULATORY COMMISSION, Standard Review Plan for the Review of Safety Analysis Reports for Nuclear Power Plants: LWR Edition (NUREG-0800, Formerly issued as NUREG-75/087), US NRC, Rockville https://www.nrc.gov/reading-rm/doc-collections/nuregs/staff/sr0800/index.html

[II–17] UNITED STATES NUCLEAR REGULATORY COMMISSION, Guidance for a Technology-Inclusive Content-of-Application Methodology to Inform the Licensing

Basis and Content of Applications for Licenses, Certifications, and Approvals for Non-Light-Water Reactors, Regulatory Guide 1.253, US NRC, Rockville (2024), https://www.nrc.gov/docs/ML2326/ML23269A222.pdf

[II–18] UNITED STATES NUCLEAR REGULATORY COMMISSION, Review of Risk-Informed, Technology-Inclusive Advanced Reactor Applications – Roadmap, Interim Staff Guidance DANU-ISG-2022-01, US NRC, Rockville (2024), https://www.nrc.gov/docs/ML2327/ML23277A139.pdf

[II–19] UNITED STATES NUCLEAR REGULATORY COMMISSION, Advanced Reactor Content of Application Project, Chapter 2, Site Information, Interim Staff Guidance DANU-ISG-2022-02, US NRC, Rockville (2023), https://www.nrc.gov/docs/ML2204/ML22048B541.pdf

[II–20] UNITED STATES NUCLEAR REGULATORY COMMISSION, Advanced Reactor Content of Application Project, Chapter 9, Control of Routine Plant Radioactive Effluents, Plant Contamination and Solid Waste, Interim Staff Guidance DANU-ISG-2022-03, US NRC, Rockville (2023), https://www.nrc.gov/docs/ML2204/ML22048B543.pdf

[II–21] UNITED STATES NUCLEAR REGULATORY COMMISSION, Advanced Reactor Content of Application Project, Chapter 10, Control of Occupational Dose, Draft Interim Staff Guidance DANU-ISG-2022-04, US NRC, Rockville (2023), https://www.nrc.gov/docs/ML2204/ML22048B544.pdf

[II–22] UNITED STATES NUCLEAR REGULATORY COMMISSION, Advanced Reactor Content of Application Project, Chapter 11, Organization and Human-System Considerations, Interim Staff Guidance DANU-ISG-2022-05, US NRC, Rockville (2023), https://www.nrc.gov/docs/ML2328/ML23283A108.pdf

[II–23] UNITED STATES NUCLEAR REGULATORY COMMISSION, Advanced Reactor Content of Application Project, Chapter 12, Post-Construction Inspection, Testing, and Analysis Program, Draft Interim Staff Guidance DANU-ISG-2022-06, US NRC, Rockville (2023), https://www.nrc.gov/docs/ML2207/ML22074A142.pdf

[II–24] UNITED STATES NUCLEAR REGULATORY COMMISSION, Advanced Reactor Content of Application Project, Risk-Informed Inservice Inspection/Inservice Testing Programs for Non-LWRs, Draft Interim Staff Guidance DANU-ISG-2022-07, US NRC, Rockville (2023), https://www.nrc.gov/docs/ML2204/ML22048B549.pdf

[II–25] UNITED STATES NUCLEAR REGULATORY COMMISSION, Advanced Reactor Content of Application Project, Risk-Informed Technical Specifications, Draft Interim Staff Guidance DANU-ISG-2022-08, US NRC, Rockville (2023), https://www.nrc.gov/docs/ML2204/ML22048B548.pdf

[II–26] UNITED STATES NUCLEAR REGULATORY COMMISSION, Advanced Reactor Content of Application Project, Risk-Informed, Performance-Based Fire Protection Program (for Operations), Interim Staff Guidance DANU-ISG-2022-09, US NRC, Rockville (2024), https://www.nrc.gov/docs/ML2327/ML23277A147.pdf

ANNEX III. REVIEW OF TYPES OF COOPERATION ON REGULATORY REVIEWS OF ADVANCED REACTORS

Regulatory bodies can cooperate during the regulatory review of advanced reactors through various approaches. These approaches are defined by specific characteristics, including the types of organization involved, the number of participating States, and the role of external bodies or networks. This annex summarizes the potential cooperation methods and the factors influencing their implementation.

Three main types of regulatory review are envisaged as part of cooperative activities for pre-licensing and licensing. These are:

- *Joint review:* a review in which a team nominated by the participating regulatory bodies, including TSOs, jointly reviews all the information against agreed requirements and comes to a joint decision. After completing a specific thematic review, each organization can cross-verify the results through a peer review process. All participants contribute to a consolidated and verified final review report. This process facilitates the discussion of problematic issues, reaching consensus on critical matters, and ensuring efficient joint reviews and verification of results. Significant opportunities for efficiencies can arise in these reviews, where one participating review team can benefit from aspects of a review conducted by another, thereby reducing time, costs and potential duplication of effort.

- *Collaborative review:* a review that involves independent assessments of information against specific national requirements. These reviews are conducted by teams nominated by participating regulatory bodies, with the process involving mutual consultation and independent decision making throughout.

  Efficiencies can be achieved by using methods and results from collaborative review teams. Participating teams can engage in discussions about the review outcomes, associated basis and any discrepancies. These discussions allow participants to adjust their reports, enabling those reviewing the same reactor design to benefit from insights from a broader pool of experts.

- *Leveraging of regulatory reviews:* reviews that involve a team, nominated by a participating regulatory body, conducting a review against its own requirements while seeking to benefit from pre-existing reviews made by other regulatory bodies. This approach allows for the integration of results from reviews conducted by others, enhancing the efficiency of the process.

  This leveraging process often begins after establishing a bilateral or multilateral agreement, which allows one participating review team to utilize results from previously completed reviews. Agreements formed after the initial review enable regulatory bodies to request specific parts of the design documentation relevant to their needs, further streamlining the review process.

These three types of approach could be combined by, for example, allocating different types to different technical areas or different participants. In addition, leveraging could be applied to many kinds of review that include ongoing or completed pre-licensing and licensing reviews (both cooperative and without cooperation). Beyond this, other factors that can influence the nature of a regulatory cooperation are expanded upon in the three sections that follow.

III–1. CHARACTERISTICS OF THE ORGANIZATIONS INVOLVED IN THE COOPERATION

The organizations involved in cooperative reviews typically include regulatory bodies, possibly more than one per State, and TSOs. They may also involve external consultants, government departments and other interested parties. The dynamics and effectiveness of the cooperation can be influenced by several factors discussed below.

III–1.1. The experience that participants bring to the project

Inexperienced regulatory bodies may join cooperative reviews with different objectives than their more established counterparts, such as prioritizing staff training. This could affect their preferred review type and the specific regulatory body they choose to cooperate with. For instance, when a country constructs its first NPP, its regulatory body may seek guidance from the regulatory body of the State that supplies the reactor. In general, and regardless of experience, all regulatory bodies can benefit from exposure to diverse methodologies and constructive challenges to their assumptions. Cooperation provides a platform for participants to fill gaps in expertise by leveraging the strengths of others.

III–1.2. Regulatory approach

Regulatory approaches range from highly prescriptive, with detailed requirements, to non-prescriptive, where high level standards are set. Proponents of prescriptive approaches argue for certainty, while those preferring non-prescriptive methods emphasize the applicant's responsibility for proving safety and the flexibility it offers.

Differences in regulatory culture can pose challenges in agreeing on common requirements, potentially making joint reviews unfeasible. Conversely, compromise and joint reviews are more feasible when participants share similar regulatory philosophies.

III–1.3. Regulatory requirements

Regulatory requirements often share high level objectives but can differ significantly in detail due to varying regulatory approaches and the prevalence of different reactor designs in each State. Agreeing on common standards requires compromise and, as with the previous item, this will be simpler when the participants have similar regulatory requirements.

III–1.4. Language

Language diversity is a common obstacle to cooperation but one that can usually be overcome with appropriate translation and communication strategies.

III–2. NUMBER OF STATES PARTICIPATING

While it is clear that a cooperation that includes many States has the potential for making maximum progress towards harmonization, successful cooperation will also become more difficult than with, say, a bilateral arrangement. This is because larger numbers will introduce greater diversity in the four factors listed above while also increasing the administrative burden. In summary, the number of participating States/regulatory bodies/TSOs could impact the implementation of the cooperation as follows:

- *Two States/regulatory bodies/TSOs review* is the simplest case of cooperation in regulatory reviews (including pre-licensing reviews). Two States/regulatory bodies/TSOs who are interested in the same design can choose to enter into a bi-lateral agreement (or it could be made under an existing agreement), which defines the process of the review and its specifics such as confidentiality issues, financial issues, what language(s) will be used for the review process, amount of information that will be available to the international community after the review is completed, etc.

- *A larger group of States/regulatory bodies/TSOs*, who are all interested in the same design, may wish to work together. In some cases, the group of States/regulatory bodies/TSOs could collaborate, potentially with assistance from an international organization since it may be challenging to reach a consensus with a large number of different regulatory regimes.

III–3. INCLUSION OF AN EXTERNAL FACILITATOR

By removing the administrative burden from the actual participants, the inclusion of an external facilitator into a cooperative review makes it easier to manage larger numbers of participants and States. This facilitator could, for example, provide project management services, a document repository, translation services, secretariat etc. Suitable candidates might be international or regional organizations such as IAEA, NEA/OECD, European Commission, Western Europe Nuclear Regulators Association, Commission of the Commonwealth of Independent States etc. The Terms of Reference of such a body will need to clearly define its functions and responsibilities, requirements for information exchange and confidentiality etc. In summary, the participation of an external body or network acting as secretariat or for administration of the regulatory cooperation could impact the implementation of the cooperation as follows:

- *Participation of an existing international organization* with sufficient experience and resources could reduce the administrative burden on regulatory bodies/TSOs.

- *Participation of existing regional network/organization* in the review process performing administrator/coordinator/overseer functions may be appropriate in case of bi-lateral or multi-lateral agreement between regulatory bodies/TSOs with similar information restrictions. Particularly, for small groups of participants it may be advisable to involve existing regional network/organization, including representatives of regulatory bodies/TSOs of a particular region (such as WENRA, CIS Commission, etc.). Using the services of such an organization could reduce the administrative burden on the cooperation participants.

- *New body/network created within the framework of bilateral or multilateral agreement* whose purpose would be to provide an administrator/coordinator/overseer function to the cooperation. Such a body/network may be established in the form of an interdepartmental commission, a working group, or another form. The Terms of Reference of such a body are expected to clearly define its main functions, area of responsibility, composition of participants, as well as requirements for information exchange and confidentiality. The creation of a new body/network will reduce the burden on regulatory bodies/TSOs in terms of organizing the process of cooperation in regulatory reviews, will ensure the exchange of information, taking into account all the existing restrictions on the transfer of information, to limit the uncertainty of the influence of such a body/network on the review process.

- *Only the regulatory bodies/TSO parties to the cooperation.* In general, participation of a body/network as secretariat is not obligatory. If all interested parties are willing, then it is possible to establish one of the regulatory bodies/TSO as administrator/coordinator/overseer of the review process, subject to written agreement of all interested parties.

LIST OF ABBREVIATIONS

ASN	Autorité de Sûreté Nucléaire
ASNR	Authority for Nuclear Safety and Radiation Protection
CFR	Code of Federal Regulations
CNSC	Canadian Nuclear Safety Commission
CSA	comprehensive safeguards agreement
DIO	design information owner
DSA	deterministic safety analysis
DSWG	design specific working group
EU	European Union
EDF	Électricité de France S.A.
ENSREG	European Nuclear Safety Regulators Group
FANR	Federal Authority for Nuclear Regulation
FOAK	first of a kind
FSAR	final safety analysis report
iDAC	Interim Design Acceptance Confirmation
IRSN	Institute for Radiological Protection and Nuclear Safety
ISWG	issue specific working group
JER	Joint Early Review
KINAC	Korea Institute of Nuclear Non-proliferation and Control
KINS	Korea Institute of Nuclear Safety
LPSAR	licensing preliminary safety analysis report
LWR	light water reactor
MDEP	Multinational Design Evaluation Programme
MEST	Ministry of Education, Science and Technology
MoC	memorandum of cooperation
MoU	memorandum of understanding
NEA/OECD	Nuclear Energy Agency of the Organisation for Economic Co-operation and Development
NDA	non-disclosure agreement
NGO	non-governmental organization
NHSI	Nuclear Harmonization and Standardization Initiative
NPP	nuclear power plant
NSSC	Nuclear Safety and Security Commission
NSR	nuclear security related
ONR	Office for Nuclear Regulation
PSA	probabilistic safety assessment
PSAR	preliminary safety analysis report
RAI	request for additional information
RFI	request for information
ROK	Republic of Korea
SAR	safety analysis report
SBD	safeguards by design
SER	safety evaluation report
SMR	small modular reactor

SSC	structure, system and component
SSM	Swedish Radiation Safety Authority
STUK	Radiation and Nuclear Safety Authority
SÚJB	State Office for Nuclear Safety
SÚRO	National Radiation Protection Institute
ToR	terms of reference
TSO	technical support organization
UAE	United Arab Emirates
US NRC	United States Nuclear Regulatory Commission

CONTRIBUTORS TO DRAFTING AND REVIEW

Akbay, B.	Nuclear Regulatory Authority, Türkiye
Alamsyah, R.	Nuclear Energy Regulatory Agency, Indonesia
Alm-Lytz, K. M.	International Atomic Energy Agency
Antonsson, A.	Swedish Radiation Safety Authority, Sweden
Arias, M.	Nuclear Regulatory Authority, Argentina
Ascic, M.	International Atomic Energy Agency
Becker, G.	NuScale Power, United States of America
Belyea, S.	Canadian Nuclear Safety Commission, Canada
Benc, A.	Nuclear Regulatory Authority, Slovakia
Brandão Nogueira Vidal, R.	National Nuclear Energy Commission, Brazil
Calle Vives, P.	International Atomic Energy Agency
Carson, A.	World Nuclear Association, United Kingdom
Chernyakhovskaya, Y.	Rosatom Service Joint Stock Company, Russian Federation
Choren, A.	Rosca Solutions, United States of America
Coenen, S.	Federal Agency for Nuclear Control, Belgium
Crossland, I.	Consultant, United Kingdom
Daubresse, O.	Électricité de France, France
Domínguez, C.	Nuclear Regulatory Authority, Argentina
Duman Yogurtoglu, L. N.	Nuclear Regulatory Authority, Türkiye
Fosaaen, C.	NuScale Power, United States of America
Grigroyan, V.	Armenian Nuclear Regulatory Authority, Armenia
Gunawan, I.	Nuclear Energy Regulatory Agency, Indonesia
Herbach, J. D.	International Atomic Energy Agency
Howlett, C.	Canadian Nuclear Safety Commission, Canada
Irawan, D.	Permanent Mission, Indonesia
Janzen, E.	Canadian Nuclear Safety Commission, Canada
Jo, S. Y.	Korea Institute of Nuclear Nonproliferation and Control, Republic of Korea
Kennedy, W.	Nuclear Regulatory Commission, United States of America
Keppen, H.	NuScale Power, United States of America
Kim, G.	Nuclear Regulatory Commission, United States of America
Kim, J. Y.	Nuclear Safety and Security Commission, Republic of Korea
Kirley, R.	Office for Nuclear Regulation, United Kingdom
Kulkarni, P.	Atomic Energy Regulatory Board, India
Künzel, K.	State Office for Nuclear Safety, Czech Republic
Kuryndin, A.	Scientific and Engineering Centre for Nuclear and Radiation Safety, Russian Federation
Lázi, D.	Hungarian Atomic Energy Authority, Hungary
Lilja, E.	Swedish Radiation Safety Authority, Sweden
Lim, S. G.	Korea Hydro & Nuclear Power, Republic of Korea
Ma, G.	National Security Council, China
MacInnes, K.	Office for Nuclear Regulation, United Kingdom
Marklund, C. R.	Swedish Radiation Safety Authority, Sweden
Meadors, B.	NuScale Power, United States of America
Michel, E.	Nuclear Regulatory Commission, United States of America

Miller, D.	Canadian Nuclear Safety Commission, Canada
Mistryugov, D.	Scientific and Engineering Centre for Nuclear and Radiation Safety, Russian Federation
Nagrale, D. B.	Atomic Energy Regulatory Board, India
Ni, M.	National Security Council, China
Olcese, J.	National Atomic Energy Commission, Argentina
Panker, M.	Hungarian Atomic Energy Authority, Hungary
Perinet, P.	Électricité de France, France
Piotukh, V.	International Atomic Energy Agency
Pócza, K.	Hungarian Atomic Energy Authority, Hungary
Rini, B.	International Atomic Energy Agency
Rychkov, M.	Westinghouse Electric Company, United States of America
Salisbury, M.	Rolls-Royce SMR, United Kingdom
Sato, Y.	Nuclear Regulation Authority, Japan
Savyolov, S.	Scientific and Engineering Centre for Nuclear and Radiation Safety, Russian Federation
Segala, J.	Nuclear Regulatory Commission, United States of America
Sinegribov, S.	Scientific and Engineering Centre for Nuclear and Radiation Safety, Russian Federation
Sinha, S.	Atomic Energy Regulatory Board, India
Song, C.	National Nuclear Safety Administration, China
Song, S. C.	Korea Institute of Nuclear Safety, Republic of Korea
Subrtova, N.	Nuclear Regulatory Authority, Slovakia
Temple, B.	NuScale Power, United States of America
Teranishi, K.	Nuclear Regulation Authority, Japan
Tongal Özkan, A.	Nuclear Regulatory Authority, Türkiye
Wang, J.	National Nuclear Safety Administration, China
Wetherall, A.	International Atomic Energy Agency
Yamada, T.	Hitachi-GE Nuclear Energy, Japan
Zhou, L.	Permanent Mission of the People's Republic of China to the United Nations, China
Zong, B.	State Nuclear Security Technology Center, China

CONTACT IAEA PUBLISHING

Feedback on IAEA publications may be given via the on-line form available at:
www.iaea.org/publications/feedback

This form may also be used to report safety issues or environmental queries concerning IAEA publications.

Alternatively, contact IAEA Publishing:

Publishing Section
International Atomic Energy Agency
Vienna International Centre, PO Box 100, 1400 Vienna, Austria
Telephone: +43 1 2600 22529 or 22530
Email: sales.publications@iaea.org
www.iaea.org/publications

Priced and unpriced IAEA publications may be ordered directly from the IAEA.

ORDERING LOCALLY

Priced IAEA publications may be purchased from regional distributors and from major local booksellers.

25-04271E

Printed and bound by CPI Group (UK) Ltd, Croydon, CR0 4YY

06/07/2026

02160600-0007